Explaining Chemical Change

Student Exercises and Teachers' Guide for

Grade Nine Academic Science

Jim Ross — *The University of Western Ontario*

Mike Lattner — *Algonquin and Lakeshore Catholic District School Board*

 *London Ontario Canada*

National Library of Canada Cataloguing in Publication	Ross, Jim (James William), 1952-
	Explaining chemical change : student exercises and teacher's guide for grade nine academic science / Jim Ross, Mike Lattner.
	ISBN 978-1-897007-01-3
	1. Chemistry--Study and teaching (Secondary)--Activity programs. 2. Chemistry--Study and teaching (Secondary) I. Lattner, Mike, 1957- II. Title.
	QD33.2.R68 2004 540'.71'2 C2004-902561-9
Authors	Jim Ross Mike Lattner
Printer	CreateSpace
Cover Design	Images, London, Ontario Canada

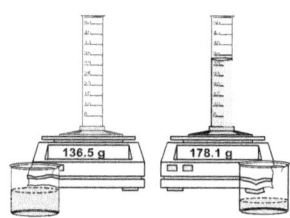

© Copyright 2004 by Ross Lattner Educational Consultants.

All rights reserved. The use of any part of this publication, reproduced, transmitted in any form or by any means, electronic, mechanical, photocopying, recording or otherwise, or stored in a retrieval system, without the prior consent of the publisher, is an infringement of the copyright law and is forbidden.

Permission is granted to the individual teacher who purchases one copy of *Explaining Chemical Change*, to reproduce the student activities for use in his / her classroom only. Reproduction of these materials for an entire school, or for a school system or for other colleagues or for commercial sale is strictly prohibited.

ISBN	978-1-897007-01-3
Offices	London Ontario Canada

To teachers, parents and students everywhere who desire to bring about new ways of understanding the world.

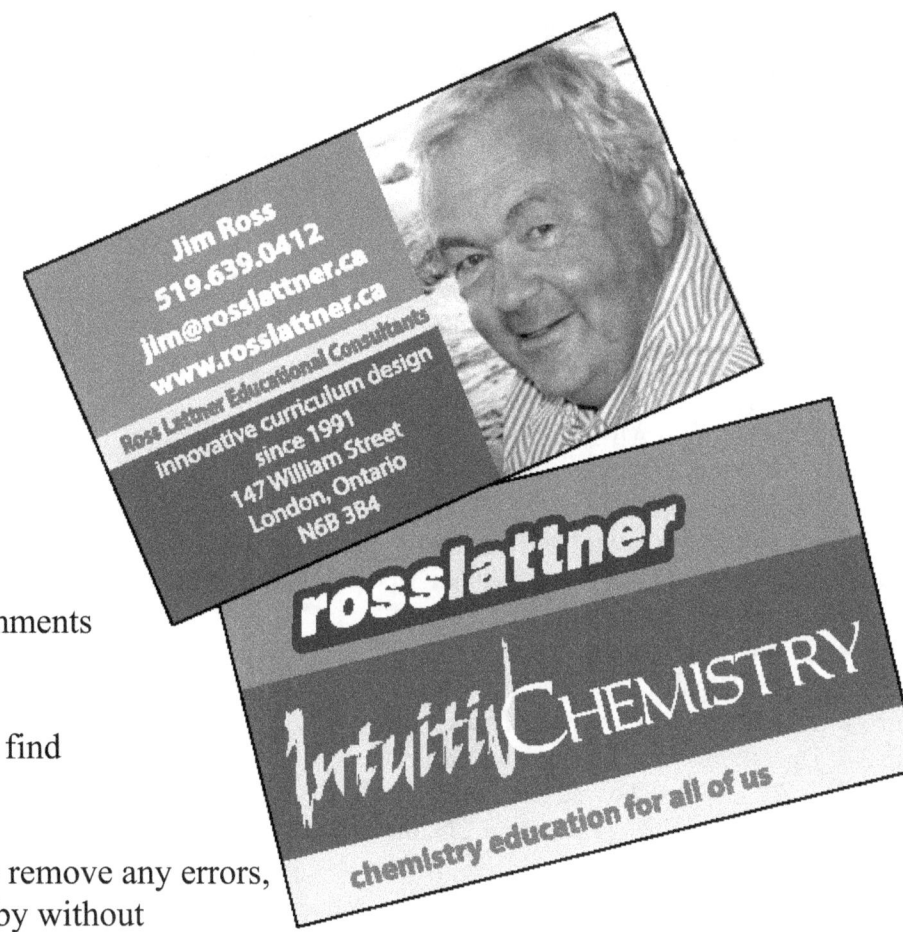

We welcome your comments and suggestions.

Let us know what you find most useful.

We've worked hard to remove any errors, but don't let a day go by without letting us know if you find one.

Stay in touch.

Jim Ross

Our thanks to all of the wonderful people at the Faculty of Education, the University of Western Ontario.

Special thanks to Willie O'Donnell, a man who thinks of the students before all else.

Explaining Chemical Change
Table of Contents

1 Teaching Chemical Change 1

Unit Planning Notes: 2
Lab 1.1: Particles and Physical Change 4
Lab 1.2: Sublimation of Iodine 4
Lab 1.3: Classifying Elements as Metals and Non-metals 6
Lab 1.4: Classifying Substances by Composition 8
Lab 1.5: The Electrolysis of Water 10
Quiz 1.6: Dalton's Theory 12
Lab 2.1: The Bohr-Rutherford Model of the Atom 14
Activity 2.2: Organization of the Periodic Table 16
Activity 2.3: Bohr, Rutherford and the Periodic Table 16
Activity 2.5: Atomic Radius and the Periodic Table 20
Activity 2.6: Putting it All Together 22
Resource 2.7: The Standard Periodic Table 24
Quiz 2.8: Periodic Table 24
Lab 3.1: Reactions of Group 17, Halogens 26
Lab 3.2: Properties of Group 1, the Alkali Metals 28
Lab 3.3: Reactions of the Row Three Elements 30
Lab 3.4: Elements and Compounds 32
Lab 3.5: Ionic Crystals, Covalent Molecules and Networks 34
Lab 3.6: Reactions of Acids with Metals 36
Lab 3.7: Decomposition of a Covalent Molecule: Hydrogen Peroxide 38
Lab 3.8: Is All Carbon Dioxide the Same? 40
Lab 3.9: Does Mass Change During a Chemical Reaction? 42

Explaining Chemical Change
Table of Contents

2 Explaining Chemical Change 45

Introduction: Three Theories of Chemical Change 46
Lab 1.1: Particles and Physical Change 48
Lab 1.2: Sublimation of Iodine 50
Lab 1.3: Classifying Elements as Metals and Non-metals 52
Lab 1.4: Classifying Substances by Composition 54
Lab 1.5: The Electrolysis of Water 56
Quiz 1.6: Dalton's Theory 58
Lab 2.1: The Bohr-Rutherford Model of the Atom 62
Activity 2.2: Organization of the Periodic Table 64
Activity 2.3: Bohr, Rutherford and the Periodic Table 66
Activity 2.5: Atomic Radius and the Periodic Table 70
Activity 2.6: Putting it All Together 72
Resource 2.7: The Standard Periodic Table 74
Quiz 2.8: Periodic Table 75
Lab 3.1: Reactions of Group 17, the Halogens 78
Lab 3.2: Properties of Group 1, the Alkali Metals 80
Lab 3.3: Reactions of the Row Three Elements 82
Lab 3.4: Elements and Compounds 84
Lab 3.5: Ionic Crystals, Covalent Molecules, and Networks 86
Lab 3.6: Reaction of Acids with Metals 88
Lab 3.7: Decomposition of a Covalent Molecule: Hydrogen Peroxide 90
Lab 3.8: Is All Carbon Dioxide the Same? 92
Lab 3.9: Does Mass Change During a Chemical Reaction? 94
Quiz 3.10: Chemical Change 96

Appendix: Laboratory Safety 100

Explaining Chemical Change

1 Teaching Chemical Change

Title: Explaining Chemical Change

Time Allocation: 27.5 hours (22 periods of 75 minutes each)

Authors: Jim Ross and Mike Lattner

Date: May 2003

Unit Description: An introduction to the periodic table, this unit starts from a discussion of the composition of matter from sub-atomic particles to solutions and mixtures. A brief survey of the s-block and p-block elements is included. The relationship between atomic structure, chemical behaviour, and organization of the periodic table is explored.

The unit itself is subdivided into three major sections. Each section will take a little more than one week to complete.

1. Chemical composition of matter. Physical change, chemical change, elements and compounds

2. The periodic table, and its connections to both chemical behaviour and atomic structure.

3. Chemical reactions chosen to highlight the relationship between the periodic table, and to demonstrate the four common classes of chemical reactions.

At the end of each section is a thorough quiz

Strand: Chemistry

Expectations: Overall Expectations: CHV 01 - 03
Specific Expectations: CH1.01 - .15; CH2.01 - .06

Explaining Chemical Change

9 Academic Science Teachers' Guide

> *It is much easier to learn how to use a small number of dynamic explanatory propositions than to memorize a vast number of specific "facts".*

Unit Planning Notes: This unit proceeds from the previous unit on electricity. Ultimately, all chemical phenomena are manifestations of the electrical force. The educational challenge, then, is to encourage the student to deal with chemistry from the most unifying point of view, that is, from the perspective of the electrical constituents of matter. The alternative is to ask the student to deal with an infinity of instances. Therefore, this unit emphasizes the following:

The Particle Theory is absolutely fundamental, as science is not possible without it. We cannot assume that students know it.

The Electrical Components of the Atom are the proton, the neutron, and the electron. It is not enough to be aware of them. Students must use these concepts to make decisions about chemical change. These components are organized in the Bohr - Rutherford model of the atom.

The Periodic Table is more readily learned as a systematic organization of the Bohr - Rutherford model than as a compilation of empirically observed properties of elements.

Chemical Behaviour of metals and non-metals that can be meaningfully related to the electrical composition of the atom.

Even with this structure, the chief pedagogical problem of chemistry remains the sheer number of concepts included. For learning's sake, it is important to keep the number of concepts down to human scale.

Prior Knowledge Required Very little prior knowledge is expected of the student. Limited everyday experience of metal and non-metal behaviour, some knowledge of the electrical force are assumed. Some experience with the particle theory is hoped for.

> *Mendeleev possessed encyclopedic knowledge of chemical behaviour when he "discovered" the periodic table. Can we expect a teenager to do the same?*

Teaching and Learning Strategies While the history of chemistry is important, it may not be the case that recapitulation of its historical development is the most meaningful way for a teenager to learn the discipline. Indeed, we wish to avoid the use of unnecessary categories or concepts. Instead, we wish to promote the student's use of commonly accepted theory as the basis for prediction and explanation. We chose experiments whose outcomes can be both predicted and explained by resorting to the theories previously discussed.

> *Can a teacher evaluate understanding? Or do we evaluate student application, and infer understanding?*

Assessment and Evaluation A variety of strategies are available. Day to day assessment of knowledge can follow the quizzes and the PEOE box diagrams. Clarity of communication can be assessed in the student's written explanations. Societal issues are best left to projects, which permit the student more time to reflect.

The Composition of Matter

Science and Pedagogy

The Particle Theory of Matter.
First encountered formally in Grade 7, this theory is foundational. It cannot receive too much exposure.

An average human can keep track of no more than 5 - 7 mental constructs at one time.

1. The absence of matter is a pure vacuum.
2. All matter is made of tiny particles.
3. All particles of one substance are identical.
4. The spaces between particles are small in solids and in liquids, and large in gases.
5. All particles are attracted to each other by forces.
6. Particles are in constant motion.

Dalton's Theory of Chemical Change

Students should use these theoretical propositions to explain the phenomena they encounter in this unit.

1. An element consists of only one kind of atom.
2. The smallest particle of a pure element is an atom.
3. Atoms cannot be created or destroyed in a chemical change.
4. Compounds consist of two or more different kinds of atoms.
5. The smallest particle of a pure compound is a molecule.
6. In a chemical change, the atoms are rearranged to form new kinds out molecules.

There is always a tension between historical usage of a term, and more recent understandings. The word *molecule* historically meant the smallest particle of a substance, whether an element or a compound (e.g. a molecule of neon). Today, it appears much more useful to use the word *molecule* to designate two or more atoms bonded together, and to use the word *atom* to designate single particles of an element.

The Bohr - Rutherford Model of the Atom.

All truth in science is tentative and conditional. The value of a good theory, then, is not that the theory claims to be absolutely true, as much as it claims to be useful.

1. Electrons: negative   tiny mass   around nucleus
   Protons: positive   massive   inside nucleus
   Neutrons: neutral   massive   inside nucleus
2. The nucleus contains all of the + charge (protons), and almost all of the mass (protons, neutrons) of the atom.
3. Electrons orbit the nucleus at fixed energy levels.
4. Each energy level can only hold up to 8 electrons (transition metals are exceptions).
5. Only the outermost electrons are involved in chemical change.

It is our intent to convey the essential ideas, without requiring the student to use them in ways which flatly contradict future historical developments in quantum mechanics, such as the orbital concept.

Explaining Chemical Change

9 Academic Science Teachers' Guide

Lab 1.1: Particles and Physical Change
Lab 1.2: Sublimation of Iodine

Perhaps some curriculum guidelines give insufficient attention to this absolutely fundamental scientific theory.

Learning Expectations: CH1.02 Describe the particle theory of matter, and use it to explain phenomena

To explain a phenomenon is to talk about it in terms of more fundamental, and therefore more universal, categories.

Pedagogical Issues The literature on student conceptions of science suggests that students tend to believe that matter is continuous, and not particulate. Furthermore, when students do adopt a particulate model of matter, they tend to assume that the particles are precisely like the bulk matter they constitute. Students frequently describe molecules of water as wet, and able to freeze or melt. They might think that particles of para-dichlorobenzene are white and smelly, and atoms of copper are reddish and shiny. This is simply not the case. The particles of water, para-dichlorobenzene and copper are electrical objects whose bulk properties appear to our senses only when billions of them are assembled together. It is worth spending some time to work with students on this question.

Another issue is the misapplication of everyday terms in the scientific context. For example, Iodine cannot be said to be *melting* or *boiling* in this experiment, as the temperature of the water bath is below the m.p. and b.p. of iodine.

Can you, as a science teacher, articulate and defend your own philosophy of science? What does science do? What counts as evidence?

Science Issues The main activity of science is to explain. In this exercise, some of the bulk properties of para-dichlorobenzene and iodine must be explained in terms of the speeds of the particles, and the subsequent distances and forces between the particles. Thus, iodine sublimes because as the particles speed up, they become farther apart and the forces between them ultimately become to weak to hold the rapidly moving particles together.

The Composition of Matter

Science and Pedagogy

Students must come prepared to do the experiments. At a minimum, they must complete the *Predict* and *Explain* boxes for each experiment, and read the *What Are We Doing* and *What Are We Thinking About* sections.

The Learning Activity Two experiments are included in this set, and either one or both can be completed in 20 minutes. Before the experiment
  Predict: the behaviour of the solids
  Explain: why you believe your prediction.
Pre-lab discussion can sharpen the issues for some students. After the experiment
  Observe and make records of the physical behaviour
  Explain: was your prediction correct? How can you explain what you observed in terms of the particle theory?
Allow plenty of time for discussion.

Crystalline sulfur is arranged into molecules of 8 sulfur atoms. At low temperatures, sulfur melts into a pale yellow liquid, molecules intact.

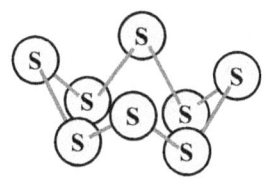

Equipment, Preparation and Resources

12 test tubes containing 5 g of para-dichlorobenzene
12 test tubes containing 0.2 g of iodine (a few crystals)
24 cork stoppers. Place these loosely in the test tubes.
12 hot plates   12 beakers

Extension and demonstration

1. Place about 5 g of sulfur in a test tube, and clamp it securely in a hand-held test tube clamp.
2. Heat the sulfur gently for a few minutes. It will melt to a clear, yellow liquid at 112°C.
3. Heat the sulfur more strongly. It will change to a dark red liquid, and become tarry and thick as the sulfur rings break into strings, and the strings bond to each other and get longer and more tangled.
4. Further heating causes the tarry liquid to become less viscous, and finally to boil.
5. Quickly pour the black, boiling liquid into a beaker of very cold water. As the tangled strings of sulfur "freeze into place" they have a tangled, stringy arrangement of particles that resembles that of rubber.
6. Take the cooled sulfur out of the water. How does it behave?
7. Leave the sulfur overnight and check it again.

At higher temperatures, the S_8 molecule is broken, and the open rings link to form a chain. The liquid abruptly becomes dark and tarry.

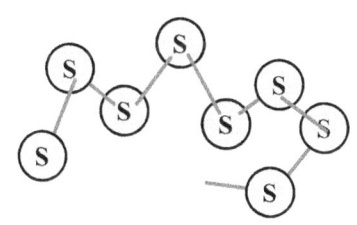

Categories: **Assessment and Evaluation**
Knowledge: Last explanation of the PEOE box diagram
Inquiry: Focus Question, PEOE cycle
Communication: Clarity and quality of the student explanation
Applications, Extensions: Where is para-dichlorobenzene used? Is that stuff safe?

Explaining Chemical Change

9 Academic Science Teachers' Guide

Lab 1.3: Classifying Elements as Metals and Non-metals

A category is a kind of cognitive bin. Classifying elements into categories involves sorting them into two or three distinct bins.

Learning Expectations: CH1.10 compare similarities and differences in properties across families of elements.

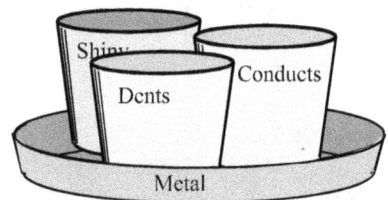

Pedagogical Issues It makes sense to build as much as possible upon the students' existing everyday categories. Students' everyday perceptual categories are helpful to sort out some of the differences between metals and non-metals. Is it *shiny* or *dull*? Does it *dent* or *shatter*? Is it a *conductor* or a *non-conductor*?

In this case, the perceptual categories lead quite directly to the scientists' categories, so students can reliably sort elements on that basis. Perceptual categories by themselves, however, have no explanatory value. The underlying electrical structure of matter must be invoked in order to explain *why* metals reflect light and *why* they can be hammered into sheets without breaking. It is ultimately these explanatory categories to which we must press.

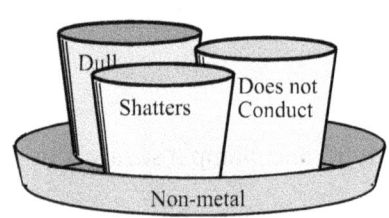

Science Issues Recall from the Electricity unit

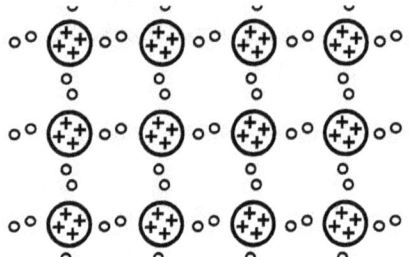

1. Non metal atoms are small and have many positive charges. They hold tightly onto electrons, and confine them to well-defined bonding pairs.
2. Metal atoms are large and have few positive charges. Their electrons move freely among atoms within the metal.
3. Mobile electrons conduct electrical charge (current).
4. Mobile electrons possess kinetic energy, and therefore conduct thermal energy rapidly through the metal.
5. Mobile electrons can oscillate in step with electromagnetic radiation, and therefore reflect light.
6. When metals are bent, the atoms just slide to other positions, held together by the extended cloud of electrons.
7. When non-metals are bent, the lattice arrangement breaks into pieces.

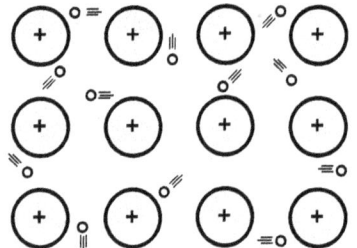

The Composition of Matter

Science and Pedagogy

The Learning Activity

Students are provided with a set of 12 elements, drawn from both metals and non-metals. Silicon, a metalloid, may also be included.

The conductivity test must be done at low voltage. A battery conductivity tester can be used, or a low voltage station can be set up, to which the students can bring their specimens.

Students can quickly categorize the elements, based upon membership in three fairly obvious categories.

Familiarity with the Classification by Composition table is helpful for the next lab (see next page).

Equipment, Preparation and Resources

12 elements must be provided, in 100 mL open beakers.

Suggested metals: copper, silver, iron, nickel, zinc, lead, tin

Suggested non-metals: Sulfur, Iodine, Oxygen, Nitrogen, Carbon (charcoal will not conduct, but graphite will)

Suggested metalloid: Silicon

Most of the metal samples will be covered with a thin oxide coating. Students can better see the properties of the metal if you first scrape the oxide off with a little steel wool or a hard knife. Sanding can score and dull the metal surface.

Categories: **Assessment and Evaluation**
Knowledge: Presented with another case, student categorizes correctly.
Inquiry:
Communication:
Applications, Extensions:

Explaining Chemical Change

9 Academic Science Teachers' Guide

We define an element as matter with one kind of atom, but we define the atom as the smallest particle of an element. How is the student supposed to get started in a circular definition?

The human process of *classification* must involve the activity of actually constructing and applying the categories to a large set of examples.

Memorizing the names of the categories cannot provide the student with a way out of the circular definition problem.

Lab 1.4: Classifying Substances by Composition

Learning Expectations: CH1.02, .03, .04, .05 Students will distinguish between atoms and molecules, compounds and elements.

Pedagogical Issues
Once again, classification by composition involves sorting substances into distinct bins. In this case, the everyday experiential categories are inadequate. Usually, we cannot simply look at a liquid and decide if it is a pure substance or a solution. This exercise is therefore quite different from the classification of elements in the last lesson. In order to successfully categorize matter by composition, some knowledge of the composition is necessary. In this exercise, the students make up the samples under study as they mix up some play dough.

Science Issues Below is a slightly expanded version of the Classification of Matter table.

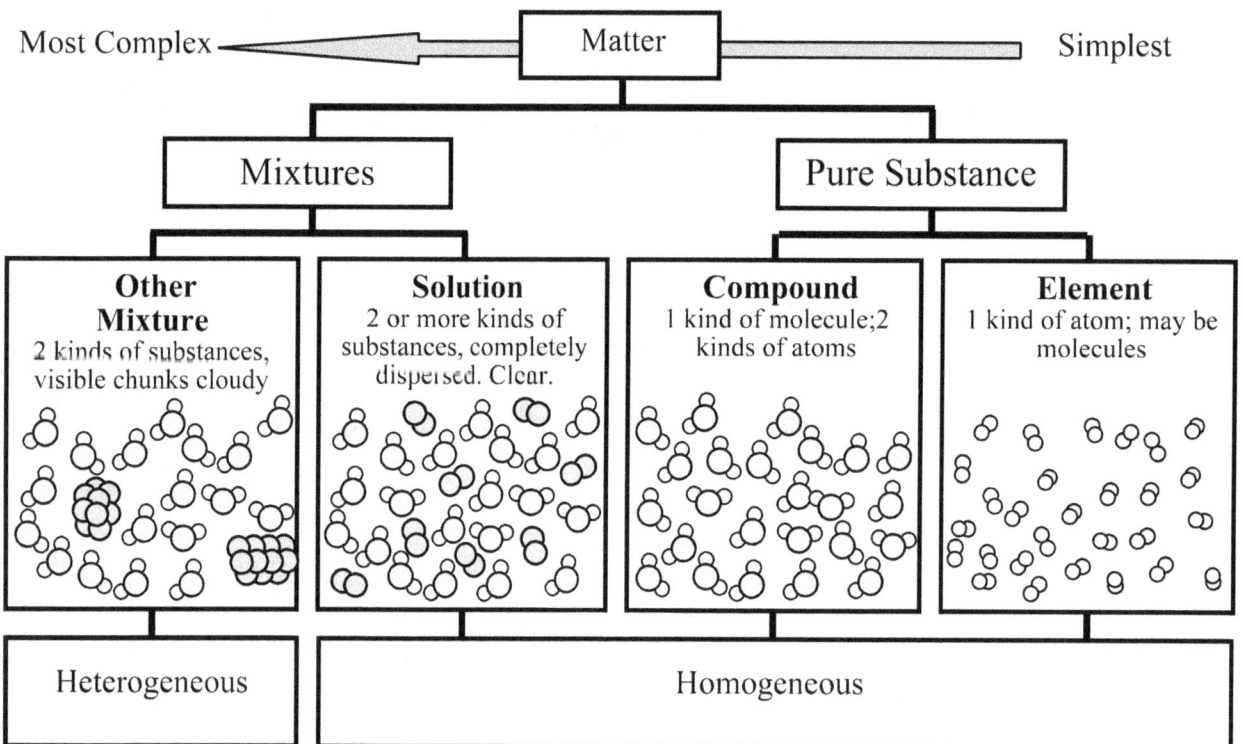

© Ross Lattner Publishing   www.rosslattner.ca

The Composition of Matter

Science and Pedagogy

Some common substances:

Substance	Formula
Aluminum Foil	Al
Ammonia	NH_3
Baking Soda	$NaHCO_3$
Bleach	NaClO
Carbon Monoxide	CO
Carbon Dioxide	CO_2
Chalk	$CaCO_3$
Charcoal	C
Chlorine	Cl_2
Copper	Cu
Diamond	C
Gasoline	C_8H_{18}
Hydrogen Peroxide	H_2O_2
Iodine	I_2
Iron	Fe
Lead	Pb
Limestone	$CaCO_3$
Natural Gas	CH_4
Nitrogen	N_2
Oxygen	O_2
Paint Remover	$(CH_3)CO(CH_3)$
Propane	C_3H_8
Rubbing Alcohol	C_3H_7OH
Salt	NaCl
Silicone	SiO_2
Sugar	$C_6H_{12}O_6$
Sulfur	S
Vinegar	CH_3COOH
Water	H_2O
Wine Alcohol	CH_3CH_2OH

The Learning Activity

The students will work together to make play dough. During the experiment, they will work with an element, some compounds, and some mixtures to make solutions and other mixtures.

The play dough itself will be used to make model atoms, colour-coded to represent a small set of elements. These can then be used to construct simple models of molecules, crystals, etc. If the models are kept sealed in plastic tubs, they will be useful for the rest of this unit.

Equipment, Preparation and Resources

Each portion of Play dough consists of the following:

500 mL water   125 mL fine salt   2 tbsp vegetable oil
2 tbsp alum   500 mL white flour
food colouring (red, green, yellow, and blue), OR
dry Tempera colour (red, green, yellow, blue, black)

One portion of each colour will be plenty.

Elements, Compounds, Atoms, and Molecules

Given the chemical formulas, students can make models of a number of common substances. To clarify the distinctions among the terms atom, element, molecule, and compound, have them sort each substance into one of the four quadrants.

	Element	Compound
Atom		
Molecule		

It should be apparent that there are no objects which are *both* Atoms and Compounds.

Categories: Assessment and Evaluation

Knowledge: Given some clues about composition, categorizes substances
Inquiry:
Communication:
Applications, Extensions:

9 Academic Science Teachers' Guide
Explaining Chemical Change

Lab 1.5: The Electrolysis of Water

An astonishing number of our conceptualizations find their roots in the everyday experience of matter moving through space. Consider love, conceptualized as motion:

> She caught his eye
> She got under his skin
> He fell in love
> She made his head spin
> He broke her heart
> They drifted apart

Human experience with objects moving in space forms the basic structure of all our subsequent meaning making.

When we teach a new way of representing nature, why not intentionally stay as close to that structure as possible?

Learning Expectations: CH1.04 Recognize compounds as substances which can be broken down by chemical means.

Pedagogical Issues Students have long found science difficult to learn. We might propose some reasons for this difficulty.

First, sentences and equations are grammatical systems, each with their own complex syntactic demands. We frequently define science concepts using sentences.

Second, the concepts, and the sentences which define them, are abstracted from the phenomena of chemistry. In the present phenomenon, we attach two wires to two electrodes, and bubbles begin to form in the water. No one ever sees any of the atoms, molecules or electrons to which the sentences and equations refer.

Finally, the sheer number of new concepts presents an obstacle. Over twenty "science concepts" are found in this lab alone. Many of these concepts have no concrete referents in everyday experience. For example, *element* and *compound* are circularly defined as opposites: what an element is, a compound is not.

Particle diagrams present a simple, powerful way through this thicket. Students can work through the chemical processes for themselves by drawing representations of atoms and molecules. The very same representations can become a shared notation, a means of communication, within the classroom.

Dalton's records contain many examples of "thinking with diagrams."

Science Issues Science is not the study of nature. It is rather the study of shared *representations* of nature. Dalton's own records show that he used a system of diagrams to represent his insights. These same diagrams became a shared notation through which he communicated his insights and engaged other scientists in debate. Even to novices, Dalton's theory of chemical change is a coherent, intuitively satisfying system when represented by particle diagrams.

The Composition of Matter

Science and Pedagogy

Teacher research question: Compare student performance on this task with their performance on the Chemical Change diagram below. In which task do your students more readily distinguish between atoms and molecules, elements and compounds?

Teacher research question. It is but a short step from this diagram to quantitative treatments of molar mass and molar volume. Would this approach work with some of your Grade 11 Chemistry students?

The Learning Activity

Before the lab, review the definitions of *element*, *compound*, *atom*, and *molecule*. This little pre-lab quiz is a learning opportunity, not an evaluation tool. The kids may get these wrong, but this assessment gives them an opportunity to improve their understanding.

The simple electrolytic cell can be set up very successfully as a demonstration. If you have enough equipment, you can run it as a class experiment.

Both of the tests should be performed on both of the gases. *When you change the current flow, will the relative volumes of the gases be changed?* Students should predict the outcome of this little modification *before* you attempt it.

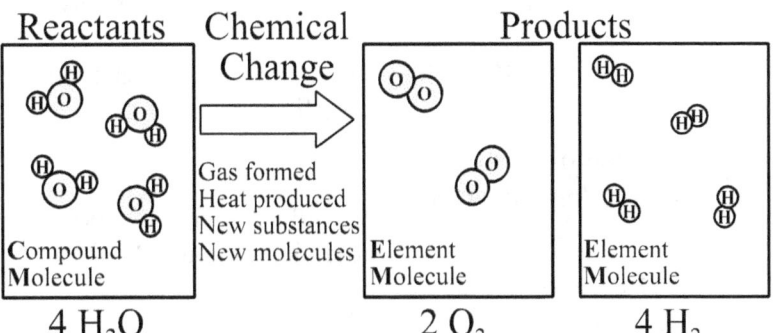

The diagram above has been completed according to instructions in the Student Exercises. The main ideas of Dalton's Theory of Chemical Change are summarized in this diagram. Students can use this kind of diagram to predict and explain the simple reactions studied in this unit.

Equipment, Preparation and Resources

The simpler the apparatus, the better. If you run it as a demonstration, try to arrange + and − in the same orientation as the students would observe in their own books.

A solution of 0.1 M Sodium Hydroxide, (4 g NaOH in a litre of water) makes a better electrolyte than an acid. In the basic environment, the electrodes are less likely to be oxidized. Oxygen is formed at very close to the ideal ratio, and metal oxides or other unwanted substances are less problematic.

Categories:
Knowledge:
Inquiry:
Communication:
Applications, Extensions:

Assessment and Evaluation
completion of the Chemical Change diagram
identification of the gases produced

9 Academic Science Teachers' Guide
Explaining Chemical Change

Quiz 1.6: Dalton's Theory

Learning Expectations: CH1.02 - .05, .10

To be useful to a student, a scientific representation must meet the following criteria:

The representation must lie within the ability of your students to produce.

The representation must solve a real scientific problem.

The representation must support student learning.

Dalton's pictures of the atom meet all three criteria. Note that it is not necessary for the representation to solve all scientific problems. Obviously, Dalton's theories cannot do that. But they can solve the problems that typical students are capable of understanding.

Pedagogical Issues

Some kind of memorization is unquestionably necessary for students to come to an understanding of chemical change. Verbatim memorization of the propositional form of Dalton's theory of chemical change is one thing, and may have its uses. Most teachers agree, however, that the ability to apply the theory to simple problems is a more desirable outcome.

These quizzes are designed to present the student with very specific problems related to Dalton's theory, and invite student analysis.

Should a student have possession of a quiz before the administration date of the test?

Why not?

As long as the students do not fill in the answers before the test, and the test is used as formative assessment of learning, possession of these quizzes motivates students with high initiative, and reassures those students with high anxiety.

In fact, the overwhelming evidence is that quick, formative assessment supports student learning.

Science Issues

The consistent use of a pictorial notation is employed here. Scientists have long invented notational systems, and developed them until they are useful means of analysis and of communication. When students become adept at using a notational system like this one, they are participating in a great scientific tradition.

The Composition of Matter

Science and Pedagogy

The Learning Activity

These quizzes can be used profitably in several ways:

Daily Pop Quiz: Did the kids do the homework? Did they understand it? You can pop one of these questions on the class the day after the lesson, and quickly assess problems.

Daily Practice Quiz: If half the class could do it on Tuesday, can they improve by Thursday?

Discussion Generator: Some questions and responses can generate controversy in the classroom. When students are required to explain their beliefs, some very fruitful learning situations can develop

Question on a later summative test: Modify and use any of these quiz items on a summative test. Students respond more confidently to structures they have seen before.

Equipment, Preparation and Resources

Grade Nine Daily quizzes.

Categories:	**Assessment and Evaluation**
Knowledge:	correct answers on quiz questions
Inquiry:	
Communication:	use of diagrams, sentences etc
Applications, Extensions:	some quiz items are new applications of knowledge

9 Academic Science Teachers' Guide
Explaining Chemical Change

Lab 2.1: The Bohr-Rutherford Model of the Atom

Learning Expectations: CH1.06, 1.07 describe the Bohr-Rutherford model of the atom, and explain how it is different from Dalton's model. Apply the Bohr - Rutherford model to the first 20 atoms

Pedagogical Issues In this lab, students are presented with a beautiful, but puzzling phenomenon: the colours produced when metal atoms are introduced into a Bunsen burner flame. In a "natural history" approach to chemistry, the students might be expected to make encyclopedic lists of elements and colours.

Colour in flames is a phenomenon in need of an explanation. The Bohr-Rutherford model can explain, at least roughly, how each metal produces its characteristic light.

Note that once the model itself is introduced (it can hardly be "discovered" by average teenagers), the explanation takes on the structure of a narrative. The production of coloured light becomes a kind of story, with a beginning, a causative chain, and an end. This structure is intuitively interesting to young students, and can become a story that they can tell others.

Science Issues

Of course this model has been superseded by Schrödinger's wave mechanics. It is nonetheless a highly fruitful model of the atom. At the introductory level of chemistry, the model has useful interpretive and predictive properties.

The existence and properties of the electron are assumed. We introduce the idea of the proton and the neutron at this point without a great deal of explanation.

> "Natural History" does not seek explanatory power.
>
> Natural history is concerned with the accumulation of, and orderly relations among, classes of natural phenomena.
>
> Young science students familiar with "natural history" approach to science tend to think of science as a body of more or less obscure and surprising facts, rather than as the discipline of explanation. But the primary purpose of science, perhaps the only purpose, is to explain the world around us.
>
> Students frequently associate high energy with danger...

The spectrum of light, and its related energy

Safe Warmth		Visible Light			Dangerous Sunburns
Low Energy					High Energy
Infra Red	Red	Green	Blue	Violet	Ultra Violet

The Periodic Table

Science and Pedagogy

An electron is flying around a nucleus at a low energy level

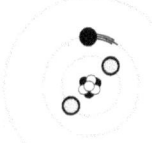

A fast moving particle in the flame collides with the atom, snapping the electron into a higher energy level.

The electron stays briefly in the high energy level.

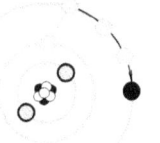

The electron snaps down to the low energy level, giving off energy in the form of light.

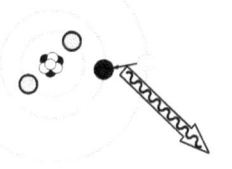

The Learning Activity

1. Allow several wooden splints to soak in each of the flasks.
2. It is important that the solutions not become contaminated. Do not mix up the wooden splints.
3. Set up, or show students how to set up, sufficient Bunsen burners. When the flames are hot blue cones,
4. Hold a soaked wooden splint in the flame, and observe the colour produced. Before the wood begins to burn, remove the splint and immerse it in the same solution.
5. Repeat as often as you need to, and make a record of your observations.

Explanation: Once the basic structure of the Bohr-Rutherford model of the atom is sketched out, the narrative at left is a simple extension of the basic model.

Equipment, Preparation and Resources

Twelve 125 mL Erlenmeyer flasks
Twelve solutions of soluble salts of any of:

Copper	sulfate, chloride, bromide
Lithium	chloride, sulfate
Sodium	chloride, sulfate
Potassium	chloride, sulfate
Strontium	chloride
Barium	chloride

Wooden splints   Bunsen burners
12 stoppers (the solutions can be kept sealed for years)

If the flame is observed through a diffraction grating, the individual colours can be separated. Several images of the flame, one for each colour, will appear in different positions, both to the left and right of the flame itself.

Nitrates, chlorates, and other oxidizers can accelerate the burning of the wooden splints. Avoid!

Categories:
Knowledge:
Inquiry:
Communication:
Applications, Extensions:

Assessment and Evaluation

last explanation of the PEOE cycle
formulation of the Focus Question
clarity of writing and diagrams

Explaining Chemical Change

9 Academic Science Teachers' Guide

Electrons fill the lowest energy levels first.

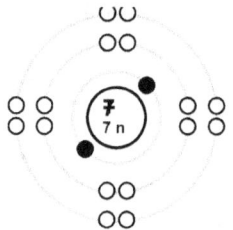

Electrons occupy the outer energy level one at a time, as singles, alone.

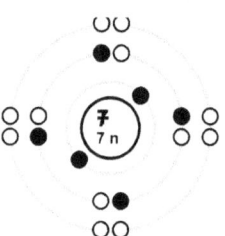

Electrons double up in pairs until that energy level is full.

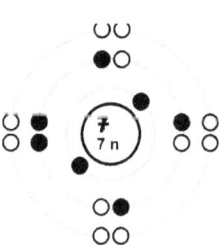

Activity 2.2: Organization of the Periodic Table
Activity 2.3: Bohr, Rutherford and the Periodic Table

Learning Expectations: CH1.07, 08 Describe the Bohr-Rutherford model, and explain its relationship to the general features of the periodic table

Pedagogical Issues The periodic table takes time to learn. This whole series provides many opportunities to work with the table. Students are challenged in these exercises to relate atomic structure, group, element name and symbol, atomic number, mass number, valence number, radius, and chemical behaviour, including metal, non-metal, and metalloid behaviour.
The purpose of the first activity is to have the student become familiar with the formal structure of the table.

The repetitious application of the Bohr-Rutherford model to the problem of the periodic table can help to do two things.
Repetition can reinforce the basic structure of the Bohr-Rutherford model, and allow students to make adjustments to their own learning.

The use of the table allows students to discern simple trends and regularities in the Bohr structure of the atom.

Science Issues The Bohr-Rutherford representation of the atom has a few subtle features that allow a smoother transition to more sophisticated models. First, we have chosen to use the term "energy level" rather than shell, because of ambiguity in the use of the word "shell" in different popular texts. Second, we have arranged the electrons in pairs, to correspond to the opposite electron spin pairings in the wave mechanical orbital model. Consistent with that model, the electron pair ⊙⊙ corresponds to a single orbital in the wave mechanical model. In accordance with the "aufbau principle" of early quantum mechanics, the electrons should be placed in the lowest energy levels first. Because of electron-electron repulsion, electrons occupy the next available orbitals in singles. Only when all singles are filled do electrons begin to double up into pairs. The atom at left shows Nitrogen's seven electrons thus arranged.

The Periodic Table

Science and Pedagogy

The Learning Activity

In groups or alone, the students complete the atomic number for each position in the table. From the atomic number in its correct position, the number of protons and electrons are determined.

The average atomic mass number indicates the average number of heavy particles in the nucleus. The only heavy particles under normal conditions are protons and neutrons. Since we know the number of protons (the atomic number), the neutrons must make up the difference between the atomic number and the mass number.

You may elect to begin a brief discussion of isotopes here.

Equipment, Preparation and Resources

Pencils, pencil crayons, coloured markers

Avoid the use of pen and permanent marker until the basic structure has been worked out.

The periodic table exercise in the student exercises

Assessment and Evaluation

Categories:
- Knowledge: response to teacher probes about elements on the table
- Inquiry: formulate a question about trends, propose an answer
- Communication:
- Applications, Extensions:

Explaining Chemical Change

9 Academic Science Teachers' Guide

The Sodium atom's complete Bohr-Rutherford structure

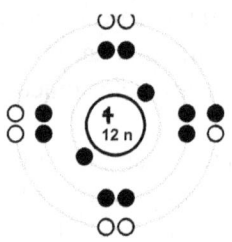

The dotted line divides the valence electrons from the atom's core.

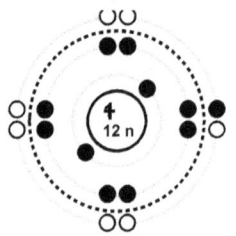

The core contains the nucleus plus completed energy levels. It resembles a noble gas.

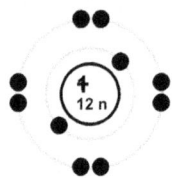

The 10− core electrons plus the 11+ nuclear charge adds up to a total core charge of 1+. The single valence electron moves around the 1+ core.

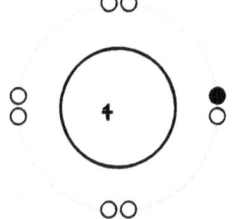

Activity 2.4: Valence Electrons, Atomic Cores and the Periodic Table

Pedagogical Issues

This exercise results in a significant reduction in the complexity of the Bohr-Rutherford model, and the introduction of two new terms. While it may be tempting to skip the previous day's work and jump directly to this simpler model, that should be avoided.

When problems or controversies arise about details of the atomic structure, the student should be able to readily resort to the original Bohr-Rutherford model.

Science Issues The atomic core consists of the nucleus plus the core electrons. The electronic configuration is always that of the previous noble gas, but the nuclear charge is always greater than that of the noble gas. The core electrons can thus be expected to behave like a noble gas, but even more strongly attracted to the nucleus, and therefore even more inert. Thus, the electrons of the core do not participate in any chemical activity.

The valence electrons and the core charge determine the gross chemical characteristics of the atoms.

The relationship between the core charge, the number of valence electrons, and the periodic table is even simpler and more direct than that of the complete Bohr-Rutherford model.

This "core-valence" diagram is immensely useful in summarizing the salient points of atomic structure. This kind of diagram is simpler, more strongly related to the periodic table, and is readily adaptable to Lewis structures of all kinds. Questions in future lab exercises ask students to explain things in terms of the core-valence diagram.

The Periodic Table

Science and Pedagogy

Core electrons are too close to the nuclear charge to be easily removed from the atom.

The Learning Activity

Introduce the students to the idea of the core electron. Core electrons are much closer to the nucleus than the valence electrons, and thus are much more strongly bound.

Introduce the students to the idea of the atomic core. The atomic core consists of the nucleus plus the core electrons. Its charge corresponds to the number of the family or group at the top of the periodic table.

The nuclear charge determines every chemical characteristic of the atom.

Introduce the student to the idea of the valence electron. Valence electrons ("fringe" electrons) are both *farther* from the nucleus and *more shielded* from the nucleus. Thus, the valence electrons can be thought of as being attracted only by the underlying atomic core charge.

Valence electrons must be differentiated from excited state electrons.

Equipment, Preparation and Resources

Are students still colouring metals, non-metals and metalloids ?

The student exercises, pencils, pens and markers are all that is required.

Categories: **Assessment and Evaluation**
Knowledge: correctly provides the core-valence diagram on demand
Inquiry: constructs questions relating the core-valence diagram and the periodic table

© Ross Lattner Publishing   www.rosslattner.ca

9 Academic Science Teachers' Guide

Explaining Chemical Change

Activity 2.5: Atomic Radius and the Periodic Table

Learning Expectations: CH1.09 Relate the Bohr-Rutherford model to properties of elements, and their position on the periodic table.

Pedagogical Issues

This activity requires the student to re-present some tabular data in the form of a graph, and then again in pictorial form.

This exercise can become the basis of building physical models of atoms out of play dough. You might wish to have students experiment with steel ball bearings, enclosed in play dough, and surrounded by small magnets. Can you reproduce the behaviour of the electrons on those atoms?

This exercise might be considered optional, in that it is mostly the re-presentation of previously existing data.

Science Issues

Atomic radius is provided in nanometers (10^{-9} m). Most students have not encountered the words picometre (10^{-12} m) or femtometre (10^{-15} m). One nanometre is one millionth of a millimetre. The wavelength of light is often expressed in nanometers (red light is 650 nm).

Even though the table would be simpler if we used 28 pm rather than 0.028 nm, it may well be the case that the concept of a nanometer is more accessible to students.

The Periodic Table

Science and Pedagogy

A person is about 2 m tall from head to toe.

one thousand times smaller:

A freckle on that person's skin is about 2 mm wide.

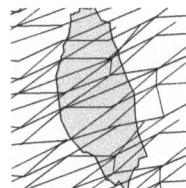

one thousand times smaller:

A skin cell in the freckle might be only 2 μm long.

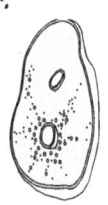

one thousand times smaller:

A cell membrane is about 2 nm, or about 50 atoms wide.

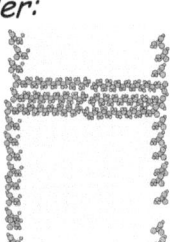

Fifty times smaller:

A single atom is a fraction of a nanometre.

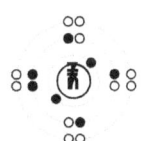

The Learning Activity

Introduce the students to the idea of atomic radius. One simple scale is in the sidebar at left.

Students will need to set a compass to a radius from a ruler. The scale is 0.01 nm : 1 mm. Once the first few atoms are done, the rest should proceed quite quickly. This exercise could easily be assigned for completion at home.

Equipment, Preparation and Resources

compass, ruler, pens, pencils, coloured markers, in addition to the student exercises.

Categories:
Knowledge:
Inquiry:
Communication:
Applications, Extensions:

Assessment and Evaluation

Questions for Later...

© Ross Lattner Publishing   21   www.rosslattner.ca

Activity 2.6: Putting it All Together

Learning Expectations: CH1.09, 10 Students compare similarities in families of elements with similarities in chemical behaviour.

Pedagogical Issues

A final opportunity to re-present the periodic trends noted so far, and to synthesize them into a coherent whole.

The intent is to provide the student with an accessible theoretical account of the structure of matter. In the next segment of the unit, students will be provided opportunities to experiment with Group 17, Group 1, and Row 3 elements. They already have some experience with metals and non-metals. If the student's experience in this course is to transcend a "natural history" of the elements, and become a "science" of explanatory power, then the student must at some point master this theoretical structure.

Science Issues

Once again we visit the metal vs. non-metal issue. Finally, we can explain basic chemical properties in terms of more fundamental electrical forces.

It may be claimed that metal and non-metal are not absolute categories of being, but rather categories of behaviour: a thing is a non-metal if it behaves like one. Because the present theoretical structure argues that non-metals have sufficient electric attraction to electrons that they can grab and hold other atoms' electrons, this account can safely support both claims as the student's knowledge grows.

The Periodic Table

Science and Pedagogy

The Learning Activity

No new terms are introduced at this point.

The students simply complete the table, and answer the questions for later. This exercise may be started in class and finished at home.

At the end, the students possess a layered account of the periodic table, from the point of view of atomic structure.

In the last cycle of the unit, this table will be used to predict and explain the chemical behaviour of some of the elements.

Equipment, Preparation and Resources

Pencils, pens, markers

Some additional probing questions:

1. How many electrons can Fluorine, Oxygen, and Nitrogen attract and hold before their valence level is filled?

2. How many electrons can Sodium, Magnesium and Aluminum lose before their valence level is empty?

3. Metalloids can either lose or gain electrons, but only reluctantly. Their electrons probably all have about the same degree of attraction toward the nucleus. Why?

Categories: **Assessment and Evaluation**
Knowledge: Quizzes
Inquiry: Labs
Communication:
Applications, Extensions:

Explaining Chemical Change

9 Academic Science Teachers' Guide

Are there such things as "generic skills" that students can learn in one context and then readily apply them to another context?

Research in learning suggests that the ability to apply a skill or concept in several unrelated contexts is itself a complex ability, and takes time and effort to learn. It may well be that the skills themselves have no inherent property of "transferability," and that the idea of "generic skills" may be misleading.

A teacher devotes classroom time and resources teaching one skill to a class. If the teacher expects the students to handily transfer it to a new context, both teacher and students are likely to be disappointed.

Resource 2.7: The Standard Periodic Table

Quiz 2.8: Periodic Table

Learning Expectations: CH1 09, 10 Students relate the Bohr-Rutherford model to properties of elements, and their position on the periodic table.

Pedagogical Issues

These tasks are sometimes more complex than they appear. Each one requires a certain level of comprehension, knowledge, and analytical process, so they are not "generic skills" tests as such.

The students' explanations for their answers may be more valuable to you as a teacher than are the answers themselves.

Science Issues

These items deal with the students' ability to manipulate various notational forms of the Bohr-Rutherford theory. They also deal with trends in the B-R representations within the periodic table. No mention has been made yet of the chemical behaviour of the elements.

The Periodic Table

The Learning Activity

These quizzes can be used profitably in several ways:

Daily Pop Quiz: Did the kids do the homework? Did they understand it? You can pop one of these questions on the class the day after the lesson, and quickly assess problems.

Daily Practice Quiz: If half the class could do it on Tuesday, can they improve by Thursday?

Discussion Generator: Some questions and responses can generate controversy in the classroom. When students are required to explain their beliefs, some very fruitful learning situations can develop.

Question on a later summative test: Use any of these quiz items on a summative test. Students respond more confidently to structures they have seen before.

Equipment, Preparation and Resources

Pencils, pens, erasers, etc.

Categories: Assessment and Evaluation
Knowledge: Correct response to the items
Inquiry:
Communication:
Applications, Extensions:

Explaining Chemical Change

The *Effort* schema is a small fragment of everyday thinking that students use to make sense of the world.

In this situation, a student might conceive of the bromine as exerting effort upon a resistive dye particle.

If this demo is to convince students that bromine is a "stronger" oxidizer, the bromine must work faster than the iodine!

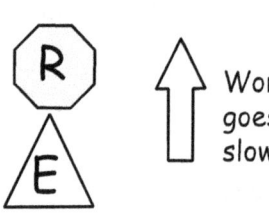

If bromine reacts faster than iodine, students mentally reorganize the phenomenon of speed into something like "bromine has a stronger attraction than iodine"

Lab 3.1: Reactions of Group 17, Halogens

Learning Expectations: CH1.09, 1.10 Students relate the Bohr-Rutherford model to properties of elements, and their position on the periodic table.

Pedagogical Issues Our teaching objective is to have students relate the small radius of the halogens to their ability to attract and hold electrons from other atoms. The smaller the radius, the more able to attract electrons. (The core and valence values are the same in the halogens). The success of this demonstration depends upon students observing greater rate of reaction with the smaller halogens. Unconsciously using *Effort* schema, students will generally associate the fastest rate with the greatest ability to attract and hold electrons. The greater the effort, the greater the results.

Science Issues The phenomenon of *bleaching* almost always involves oxidation of a coloured compound. For example, the dye molecules in food colouring have extended, conjugated double bond systems. Each double bond contains an extra pair of electrons, which an oxidizing agent such as bromine can easily remove. Once the double bond system is gone, the molecule can no longer interact with light at the particular frequency that gives the dye its characteristic colour.

While we would predict the rates chlorine > bromine > iodine in oxidative ability, in fact this is not the case in oxidation of organic dyes. Bromine is a faster bleaching agent than chlorine. Perhaps bromine's size may allow it better match to the C=C double bond. *For both safety and pedagogical reasons, don't use chlorine for the bleaching experiment. Compare bromine and iodine only.*

Because rates are very sensitive to temperature and concentration, all solutions should be around 15°C. To prepare fresh solutions at approximately equal concentrations, see instructions opposite.

A photograph provides excellent dye to bleach. Any coloured cotton or paper product will work, but more slowly. Experiment with a piece of blue denim, an old playing card, comic strips, maps, etc.

These chemicals are immediately noxious. Use the fume hood! Safety goggles at all times, and small quantities only.

Other experiments: How quickly do these solutions oxidize a ball of

Chemical Behaviour

Science and Pedagogy

The Learning Activity Teacher Demonstration. This lab contains too many chemical hazards for kids.

Before the experiment

Predict: Which halogen will be the most reactive bleaching agent... bromine, or iodine?

Explain prediction, using both valence - core diagrams, and sentences.

Label three clean dry 500 mL flasks Cl, Br, and I. Put 3 g KBr into flask Br, and 4 g KI into flask I.

Just before use, put 25 mL of bleach into each flask.

Once the solutions are prepared, place strips of a photograph in the solution, leave for a few seconds, and remove and rinse in warm water. Examine the specimens for bleaching.

After the experiment

Observe and make records of your observations.

Explain prediction, using both valence - core diagrams, and sentences.

To each flask add 450 mL of 0.1 M HCl (Hydrochloric acid).

Equipment, Preparation and Resources Solutions of halogens in water are available, but even when fresh, their concentrations are different. Once opened, their concentrations change rapidly, and there is no simple way to measure and adjust them. You can get around the problem by preparing solutions from liquid chlorine bleach, which is about 5% NaClO by mass. You need the chlorine to generate bromine and iodine. You do not need chlorine for the demonstration.

Caution: the acid will generate free chlorine gas. The Cl_2 gas will displace Br_2 and I_2.

Caution: Work in a fume hood! The resulting solutions will be 0.035 M Cl_2, Br_2 and I_2. The total volume of Cl_2 gas released will be 400 mL over a period of several hours. The Bromine will disperse more slowly, and the Iodine over a period of days.

$$2\ HCl\ +\ NaClO\ \Longrightarrow\ H_2O\ +\ NaCl\ +\ Cl_2$$

$$Cl_2\ +\ 2\ KBr\ \Longrightarrow\ 2\ KCl\ +\ Br_2$$

$$Cl_2\ +\ 2\ KI\ \Longrightarrow\ 2\ KCl\ +\ I_2$$

Categories: **Assessment and Evaluation**
Knowledge: Questions for Later
Inquiry: Generation of new questions
Communication:
Applications, Extensions: Should we use such large amounts of chlorine?

© Ross Lattner Publishing   www.rosslattner.ca

Explaining Chemical Change

9 Academic Science Teachers' Guide

When students cannot understand the forces at work in a situation, they often use the *Conflict* schema to organize the objects of their attention into "winners" and "losers." These roles have strong emotional meanings; students find them very convincing.

The struggle between agents is not symmetrical

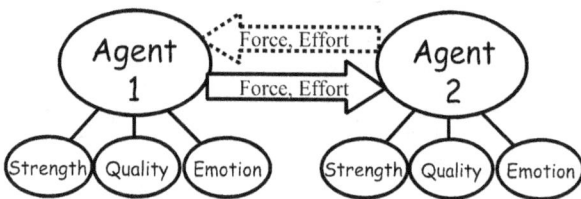

The categories of the agents are symmetrical

When water and sodium interact, students tend to mentally organize them as if they were locked in a *Conflict*.

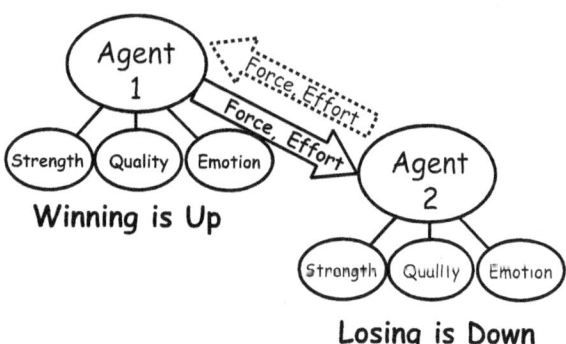

Listen for words that indicate struggle: *take over, win* or *lose, stronger* or *weaker, can't resist*. These may indicate non-scientific modes of thinking, as opposed to useful explanation.

Lab 3.2: Properties of Group 1, the Alkali Metals

Learning Expectations: CH1.09, .10
Students relate the Bohr-Rutherford model to properties of elements, and their position on the periodic table.

Pedagogical Issues Who is acting upon whom? We would like to see the students to "see" that water (hydrogen in the water, actually) can easily remove electrons from the alkali metals. The student, on the other hand, "sees" an exotic metal touching an ordinary glass of drinking water, then sparking and popping. So what is the student likely to think? That the "sodium explodes when it hits the water," and very little more than that.

One of the pedagogical tasks here is to cast the water (or the hydrogen in the water) in the role of the *agent*. It is the water that more strongly attracts the electrons of the alkali metals. That said, the problem now looks like an equal effort (water) acting upon an unequal resistance (the alkali metals).

Science Issues Once again, the key issue is safety. The metals react very quickly with water, generating hydrogen gas and often igniting it with the heat of reaction. The hydrogen can occasionally pop. Safety goggles, larger beakers, and wire gauze reduce the hazard.

As lithium reacts slowly with water, it should be the first metal the students work with.

Cut the metals carefully into 2 mm cubes, just before the experiment. Explosions occasionally happen, especially if holes are poked in the metal, or an oxide crust is permitted to form around the metal. Smooth, clean metal samples generally react smoothly.

Chemical Behaviour

Science and Pedagogy

The Learning Activity Go over all of the instructions, emphasis on clean, dry work place.

Before the experiment

It's always a good idea for the students to make their initial predictions and explanations the night before the experiment.

Predict which alkali metal will be most reactive with water
Explain the prediction, using core - valence diagrams and complete sentences

When all preparations are made (goggles on, beaker with water, tongs, wire gauze ready), the give the students small portions of lithium. Only distribute the sodium and potassium as they gain experience with handling the metals.

If a student is using natural schematic thinking, which substance will she consider to be the stronger agent: the sodium, or the water? The sodium, or the potassium? Do these notions help her learn a scientific understanding?

Students receive the metal on a petrie dish or a watch glass to take back to their desk, and handle the metal only with tongs.
As soon as they have put the metal into the water, they must cover the beaker with the wire gauze in order to prevent spattering.

After the experiment

Observe the reactivity of the alkali metals, and record.
Explain observations, using core - valence diagrams and complete sentences.

Students should use fresh water for each experiment. This is important for comparison.

Equipment, Preparation and Resources

A little phenolphthalein added to the water can make the results more colourful.

Goggles at all times. Standard student lab kit, including beakers, wire gauze, tongs, watch glass.

Handle the metals cautiously with tongs. Carefully blot up as much of the paraffin as you can. Do not touch the metal with your bare skin. Cut the metals with a sharp knife or scalpel into pieces about 2mm × 2mm × 2mm. Err on the small side.

Categories: **Assessment and Evaluation**
Knowledge: Last *Explain* box; *Questions for Later*
Inquiry:
Communication: Explanation on the Box Diagram.
Applications, Extensions:

© Ross Lattner Publishing   www.rosslattner.ca

9 Academic Science Teachers' Guide
Explaining Chemical Change

If you were an electron ●, where would you rather be?

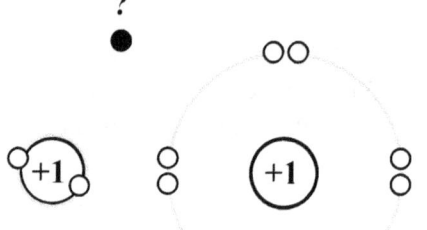

Six paces from +1 Na, or one pace from +1 H?

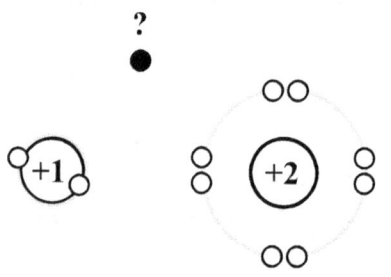

Four paces from +2 Mg or one pace from +1 H?

Lab 3.3: Reactions of the Row Three Elements

Learning Expectations: CH1.09, 1.10 Students relate the Bohr-Rutherford model to properties of elements, and their position on the periodic table.

Pedagogical Issues Two intellectual tasks are required of the student in this lab. First, the student must provide some representation of the trends in the visible chemical behaviours of the row three elements. Second, the student must provide an explanation for those phenomena.

The only samples studied in this series are the first four elements of the row. A student could be expected to see the trend among those elements, but not the remaining elements in the series.

Science Issues The explanation for the trend of activity among the first four elements should contain references to the trends in the size of the core charge, and the radius of the atoms. In general, the greater the core charge, the more tightly the electrons are held; the smaller the radius, the more tightly the electrons are held.

One of the practical problems with labs of this type is the difficulty of actually comparing oranges with oranges. Sodium, magnesium and aluminum are all metallic solids. Silicon is not entirely metallic; at least some of its hard, diamond-like character appears to be related to covalent network structure.

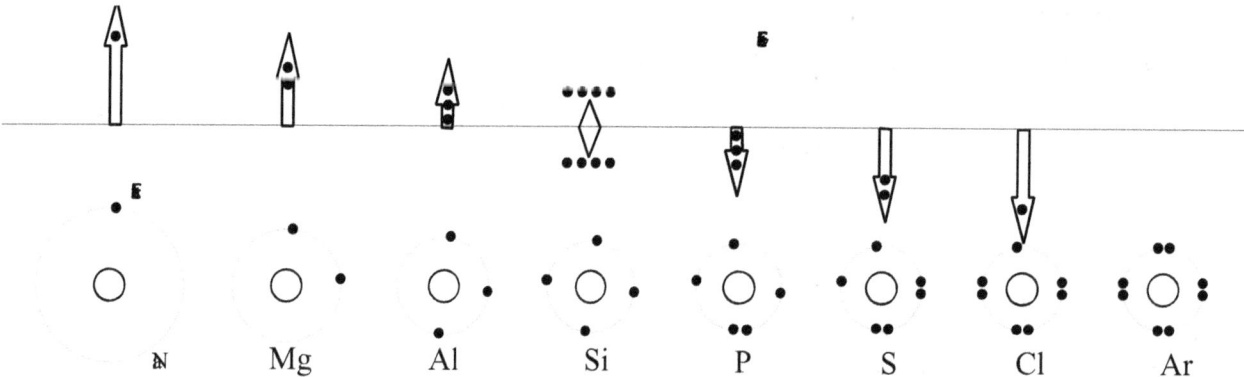

Fill in the core charge for each element. The length of the arrow indicates how readily the electrons are lost or gained. Silicon can either lose or gain electrons, but does neither readily. Argon strongly attracts its own electrons, but its valence energy level is completely full, so it cannot hold electrons from other atoms.

© Ross Lattner Publishing   www.rosslattner.ca

Chemical Behaviour

Science and Pedagogy

The Learning Activity
Students are to prepare four small 100 mL beakers with 20 mL of cold water, and have ready a wire gauze and tongs.

Before the experiment
>**Predict** the order of activity with water, from most reactive to least reactive. Then, **Predict** the order of activity with acid.
>**Explain** predictions, using the valence - core diagrams and complete sentences.

When aluminum reacts with HCl, it produces a great deal of heat. The increasing temperature can result in a fast reaction rate.

Students are given small samples of silicon, aluminum, magnesium, and sodium., which they add to the water. The Sodium reacts vigorously as before; Magnesium may show some tiny bubbles; the others appear inert. Only Mg, Al, and Si remain for the second part. The students pour off the water, and replace it with 0.5 M HCl. The Mg reacts vigorously, Al reacts slowly but releases much heat, and the Silicon reacts not at all.

After the experiment
>**Observe** the reactivity of the row three elements with water and with acid, and record your observations here
>**Explain** observations, using core - valence diagrams, and complete sentences.

The diagram at bottom of the opposite page shows how electrons would behave, relative to Silicon as the standard.

Equipment, Preparation and Resources

0.5 M HCl	Pour 40 mL of conc. HCl into 500 ml cold water. Add water to make 1 L.
Na	Cut into 2mm cubes as before
Mg	Cut Mg ribbon into 3 cm strips
Al	Strips of Al foil, about 1 cm × 3 cm.
Si	Small crystals. These are quite inert, and should be collected, washed, and kept for next year.

Categories: **Assessment and Evaluation**
Knowledge: Explanation ; Questions for Later
Inquiry:
Communication: Paragraph in Explanation
Applications, Extensions:

Explaining Chemical Change

9 Academic Science Teachers' Guide

Lab 3.4: Elements and Compounds

Elements are the simplest substances. They cannot be broken into simpler substances by chemical means. They consist of only one kind of atom.

Learning Expectations: CH1.05, .11, .14 Students depict compounds as being composed of molecules of different kinds of atoms, anticipate the chemical properties of elements based upon their position on the periodic table, and describe the visible signs of a chemical change.

Pedagogical Issues We are confronted once again with the sheer number of conceptual relationships that must be made in chemistry. The concepts *element*, *compound*, *atom* and *molecule* are re-introduced here. New understandings are possible for the student in this new context of atomic models and periodic trends. The pictorial notation of Dalton's theory is also re-introduced, to reinforce the idea of a chemical change.

Compounds consist of two or more kinds of elements. Compounds can be broken into simpler substances by chemical means. The smallest particle of a compound is a molecule. Molecules of compounds contain two or more kinds of atoms.

This lab for the first time puts different metals and non-metals together for the purpose of making a new compound. While it may be possible for a few students to read and understand simple accounts of chemical bonding themselves, it is not an expectation in this lab.

Science Issues The exact nature of the bonding arrangements is not discussed in the student exercises. Included in the lab, however, are examples of metal + non-metal reactions, and non-metal + non-metal reactions.

The covalent molecules formed, CO_2 and SO_2 are gases. It should not be much of a stretch for your students to conceptualize them as discreet molecules.

On the other hand, the ionic compounds formed, MgO and CuS, are solids. Once again, students are used to representing solids as a regular lattice of atoms, so representing an ionic lattice should be an easy step.

Caution: Sulfur dioxide is noxious. Small quantities of sulfur only, no bigger than a capital O. Good ventilation needed.

Caution: Burning Mg produces considerable radiant energy. Do not look directly at the burning magnesium. Hold a sheet of paper between the eyes and the burning magnesium.

Chemical Behaviour

Science and Pedagogy

The Learning Activity After observing the reactions you must complete each diagram to include the following:
a) Mark each box either *Element* or *Compound*. (***E*** or ***C***)
b) Mark box *Atoms*, *Molecules* or *Ionic Solid*. (***A M*** or ***I***).
c) Below the arrow, list three observable signs that chemical change has taken place.
d) Draw diagrams to show exact number of molecules.

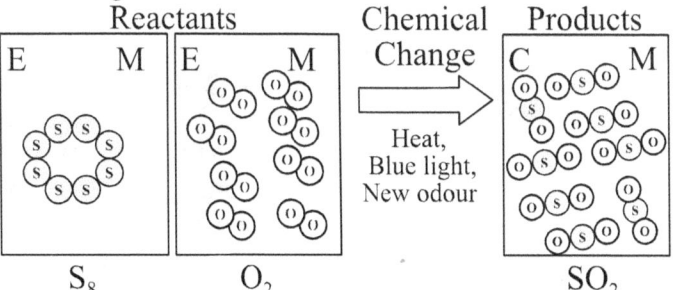

The white solid produced by magnesium is an ionic solid. Students readily comprehend the idea of a solid lattice.

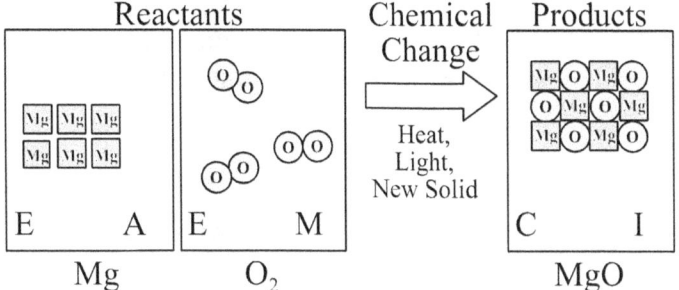

Equipment, Preparation and Resources
Sulfur, small pieces, about half a rice grain
Carbon bean sized chunks, if available
Copper clean penny
Magnesium strip, about 7 - 10 cm

Burning magnesium produces a great deal of heat and light, and if students stare at it, the retina can be burned by the infrared. Hold you're a sheet of paper between your eyes and the magnesium, so that you cannot see the flame directly.

Categories: **Assessment and Evaluation**
Knowledge: complete particle diagrams. Answers to Questions for Later
Inquiry:
Communication: use of diagrams
Applications, Extensions:

Explaining Chemical Change

9 Academic Science Teachers' Guide

Lab 3.5: Ionic Crystals, Covalent Molecules and Networks

> The "octet rule" is violated in a very large number of cases. Carbon is one of the few elements that usually follows the rule. Should we teach the octet rule?

> "Atoms!! What do they really want?"
> *Sigmund Bohr*

> Three combinations of element types can be described with this model:
>
> **Non-metal + non-metal** strongly attract each others' electrons into their own valences, forming covalent bonds.
>
> Electrons on **metal** atoms move readily to the **non-metal** atoms. The resulting oppositely charged particles are attracted to each other as ionic crystalline solids.
>
> **Metal + metal** generally form solutions, which solidify into alloys.

Learning Expectations CH1.05: Students explain how atoms combine to form ionic crystals, molecules, and covalent network solids.

Pedagogical Issues The student use of natural schematic thinking has been mentioned often. Because students rely on schemata for everyday reasoning about their environment, it is wise to avoid analogies which contain strong elements of anthropomorphism.

When students hear that sodium "wants" to lose an electron in order to "achieve" a full outer shell, or that chlorine "needs" an electron to "achieve a stable octet," they process those terms outside the realm of scientific thought. Many students cannot reconcile their own experience of wanting and needing with the rather abstract pictures on the paper, and don't understand how to proceed. What was supposed to be a helpful analogy may in fact be a substantial impediment to thinking about atoms.

An attempt is made here to use only particle analogies and the electric force, as the students extend the valence-core charge notation to include common types of chemical bonding.

Science Issues A simple description of bonding is offered here. It is probably not useful to talk about electro negativity, and percent ionic and covalent character.

It is certainly simpler, and I believe more productive, to accept that the properties which define metal and non-metal behaviour in bonding situations are precisely the same properties which cause the gross differences in appearance between metals and non-metals.

Thus, if element A looks like a metal, feels like a metal, and bends like a metal, it's probably a metal. Likewise, element B looks, feels and behaves like a non-metal. These two elements will probably form an ionic compound, whether or not their electro negativity differs by 1.4

Chemical Behaviour

Science and Pedagogy

Solubility was specifically left out because of the large number of counter-examples, the insoluble ionic solids.

Ionic crystal lattice:

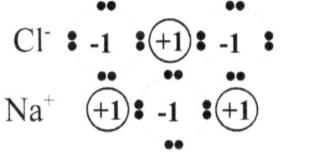

Distinct covalent molecules:

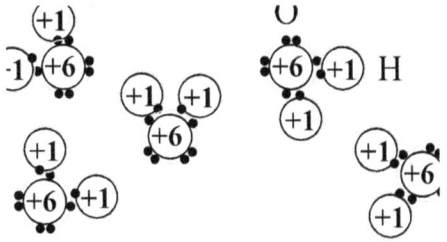

Covalent Network Solid:

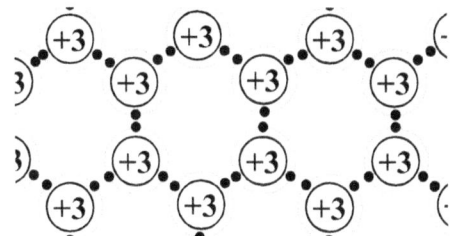

The Learning Activity

Part One: Students examine the behaviour of a small number of typical substances.

If the students put all of the solids on a cold hot plate, and then plug the hot plate in, an indication of melting point can be obtained in a few minutes.

Only small, weakly bound molecules can vaporize readily. Odour responses are typically provoked by small molecules.

Part Two: Draw representations of bonding in terms of valence electrons and core charges in all three situations. The exercise offered here is pretty standard.

Ionic Solids: It helps to show which electrons are moving, the formation of ions, and the resulting forces.

Covalent Molecules: the valence electrons are attracted by both atoms, but each atom has room for only one or two of the other's electrons.

Covalent Network Solids: the covalent bonds are exactly the same as covalent molecules, but the atoms have room for more of the others' electrons. Common examples are based upon boron, silicon and carbon, all of which have room for three or four electrons from other atoms.

Equipment, Preparation and Resources

12 Hot plates (must be cold when started)
2 Covalent Compounds:
    menthol, paraffin wax, camphor
2 Ionic Solids
    Sodium chloride, any oxide, not potassium iodide.. Use small lab grade crystals. Rock salt contains pockets of water, and can explode like popcorn.
1 Network Solid
    Silicon, graphite, diamond
2 unknowns
    Any of the above

Categories:
    Knowledge:
    Inquiry:
    Communication:
    Applications, Extensions:

Assessment and Evaluation

identifies ionic solids and covalent compounds on sight

9 Academic Science Teachers' Guide
Explaining Chemical Change

Lab 3.6: Reactions of Acids with Metals

Learning Expectations: CH2.03, .04, .05 Students conduct experiments in the behaviours of common substances, and correctly represent the chemical formulas of the substances.

The test for oxygen:
Collect a test tube of the gas
Prepare a glowing splint
Thrust the glowing splint into the test tube of gas.

If the ember bursts into flame *then* the gas is rich in oxygen.

Pedagogical Issues Introduced here is the idea of interpreting the results of a chemical test, and then using our interpretation as evidence in a larger argument. This idea, so familiar to science teachers, is by no means obvious. Indeed, we often find it difficult to see why students find it a problem at all. Let's go through this very typical test.

1. *We collect a representative sample of gas.* In this lab, the gas is collected over the acid. Is the gas pure? The test actually works best if the gas is a mixture of about 20% Hydrogen and 80% air!

The test for hydrogen:
Collect a test tube of the gas.
Prepare a flaming splint
Bring the flame to the mouth of the test tube.

If the gas explodes, *then* the gas is likely hydrogen.

2. *We try to initiate a chemical reaction.* The only way to *test* the chemical properties of a substance is to *observe its chemical changes*. In this test, the rapid oxidation of hydrogen releases a characteristically large amount of energy.

3. *We interpret the reaction observed.* In this case, we must have had some suspicion that the gas was hydrogen before we started the test, since many other gases also react explosively with air.

This may seem terribly pedantic, but too many students and teachers leap right over points 2 and 3. The chemical reaction which constitutes the test is poorly understood, so interpretation is compromised. The result is that, for many students "bang means hydrogen" (discussion continued page 38)

Is it possible to assess student understanding? Or can we only assess the *quality of student representations?*

Science Issues In keeping with the goal of deepening students' mastery of the pictorial notation of Dalton's theory of chemical change, this lab exercise produces both a molecular element, Hydrogen, and an ionic solid, Zinc Chloride. The pictorial notation we have adopted is sufficient for students to make reasonable predictions, and explain their observations in this lab.

Chemical Behaviour

Science and Pedagogy

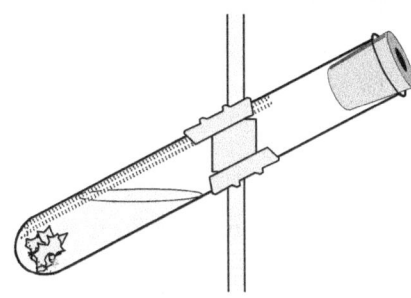

The Learning Activity Before the experiment, the students:
  Predict whether the gas produced is *Hydrogen* or *Oxygen*?
  Explain why they believe their prediction.
Then the students:
1. Clamp a test tube securely to the retort rod at a low angle, and add about 10 mL of 0.50 M hydrochloric acid.
2. Feel the temperature of the test tube before and after.
3. Gently slide one piece of zinc down the test tube into the acid, and place a one-holed stopper loosely into the mouth of the test tube. The single hole allows pressure relief.
4. Test the gas for both oxygen and hydrogen.

After the experiment
  Observe, and record their observations.
  Explain their observations, using Dalton's particle diagrams and complete sentences.
Finally, to examine the other product, students:
5. Pour the liquid through a filter, and leave the solution in a petrie dish overnight. They must observe the residue in the next class.

If all of the students dispose of their solutions through a filter at the front of the class, you can do this part as a class demonstration.

The test for hydrogen... Does it resemble a story?

Explain this chemical test:

Is that black stuff really copper? When the reaction is over, add some nitric acid, or even sodium nitrate. A blue colour can indicate the presence of copper.

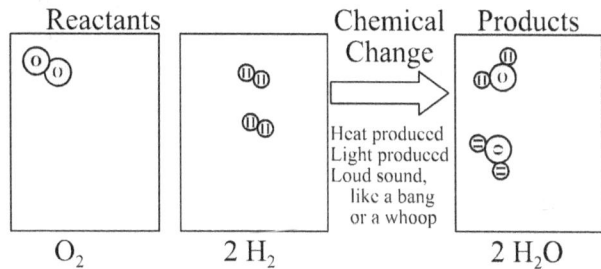

Equipment, Preparation and Resources

The acid must be prepared in advance.:
0.50 M HCl Add 40 mL of conc. HCl to 500 mL of cold water. Add more water to make 1.0 L.

The metal must be mossy zinc. This material comes in lumps, and contains a small percentage of copper. The copper speeds up the reaction, but is left behind as a black residue.

Categories:
Knowledge:
Inquiry:
Communication:
Applications, Extensions:

Assessment and Evaluation

student explanations, including their diagram, in Box 4

quality of the student explanations, including their diagram

Explaining Chemical Change

9 Academic Science Teachers' Guide

Lab 3.7: Decomposition of a Covalent Molecule: Hydrogen Peroxide

Learning Expectations: CH2.03, .04, .05
Students conduct experiments in the behaviours of common substances, and correctly represent the chemical formulas of the substances.

Pedagogical Issues (Continued from page 36) The word "evidence" has a conditional status to a scientist that it lacks in the everyday world of the student. One of the cardinal differences between the two is the *interpretation* of the observed chemical reaction. To a scientist, interpretation is the line of reasoning that links the *consequences* of a chemical reaction, to the *conditions* which made the reaction possible.

An average student can start by pictorially representing this lab in the notation of Dalton's theory. In turn, the pictorial representation supports a kind of narrative, a story which unfolds in a familiar and particular way.

In this story, the glowing ember (red hot carbon) is immersed in oxygen. Oxygen reacts with the carbon, producing more heat. No other common gas is able to react with carbon in that way.

The next time Jane puts a glowing splint into a gas, and the splint bursts into flames, she has a story in which the hidden (but suspected) identity of one of the characters is revealed.

Note that this line of reasoning is not likely, perhaps not even possible, if the student is not intimately familiar with the story.

Science Issues Tell me a story about the glowing splint...

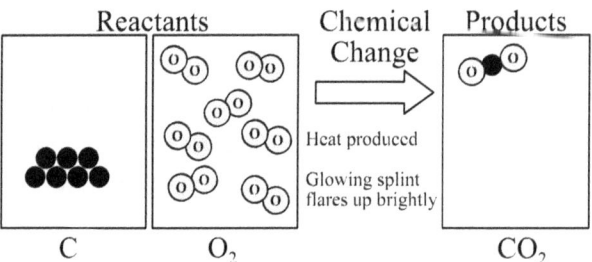

Is a chemical test a narrative? A story? When the three bears came home, and saw the pilfered porridge, they quickly made up the rest of the story.

To state the scientific method in terms of logic certainly intimidates kids who know very little of logic. More important, logic may not be the way that kids make meaning of the world.

Think of the last time you reasoned something through. Did your thinking "fill in the blanks" in a familiar story?

Or did you *really* use propositional logic?

Chemical Behaviour

Science and Pedagogy

The Learning Activity
Before the experiment, the students:
  Predict: When hydrogen peroxide decomposes, is the product *hydrogen*, or *oxygen*?
  Explain their predictions, using Dalton's theory in diagrams and sentences.

What other substances "cause" (act as a catalyst for) the decomposition of hydrogen peroxide?

As an additional experiment at home, have students try:

1. Place a tiny (half a match head) quantity of manganese dioxide into the bottom of a dry test tube. Pour 10 mL of hydrogen peroxide into it. Place a one-holed rubber stopper LOOSELY into the mouth of the test tube.

2. Test the gas for oxygen and hydrogen, and record all of your observations.

Inorganic materials such as salt, chalk, metals, sulfur.

Organic materials such as sugar, oil, alcohol, etc.

3. After the reaction has stopped, examine the manganese dioxide. Carefully pour off any remaining liquid, and replace it with fresh hydrogen peroxide. Does the manganese dioxide still have all of its "power"?

Living things such as freshly cut vegetables or plants, or small cuts or scratches on your skin.

After the experiment, the students:
  Observe and make records of the experiment.
  Explain their observations, using Dalton's theory in diagrams and sentences.

No cuts? Try a bit of fresh meat.

Equipment, Preparation and Resources Manganese dioxide is easiest to use in the form of sand-sized particles. The reaction's a little slower in that form, but it's easy to separate from the spent hydrogen peroxide.

Hydrogen Peroxide is the blonde's best friend. How does a bleach work?

Hydrogen peroxide 3% is available at pharmacies. Hydrogen peroxide decomposes easily. Contamination of almost any kind, or even just scratches inside the bottle, can accelerate decomposition. Store in the refrigerator.

Categories: **Assessment and Evaluation**
Knowledge: Correct use of the word *catalyst*
Inquiry: Home experiments
Communication:
Applications, Extensions: Is hydrogen peroxide safe? What are its uses?

Explaining Chemical Change

9 Academic Science Teachers' Guide

Lab 3.8: Is All Carbon Dioxide the Same?

Limewater must be prepared in advance.

Learning Expectations: CH2.03, .04
Students conduct experiments in the behaviours of common substances, and correctly represent the chemical formulas of the substances.

Put several heaping spoonfuls of calcium hydroxide (slaked lime) into a clean 2.5 L acid bottle. Fill the bottle with distilled water, and agitate well.

Pedagogical Issues Do students comprehend the relationship between the particle nature of matter, and the identity of bulk substances? Vitamins from natural sources, for example, are frequently believed to be qualitatively different from manufactured vitamins.

When the undissolved $Ca(OH)_2$ has settled, decant the clear limewater off into a second clean 2.5 L acid bottle, labeled "limewater."

In this experiment, students test carbon dioxide from three different sources: decomposition of sodium bicarbonate, combustion of carbon, and respiration. Scientists believe that all molecules of carbon dioxide are identical (excepting isotopic differences). Can students articulate that belief for themselves?

The identity of all carbon dioxide is supported in two ways. The chemical test provides similar results in all three cases; and the pictorial representation is the same in all three cases.

Refill the first bottle with distilled water and more calcium hydroxide, seal it, and leave it sit. It will be ready when you use up the limewater in the second bottle.

Science Issues The common laboratory test for carbon dioxide is the chemical property of carbon dioxide which forms carbonates in the presence of water. In this case, solid calcium carbonate is formed from calcium hydroxide.

Keep both bottles well sealed: they absorb carbon dioxide from the air.

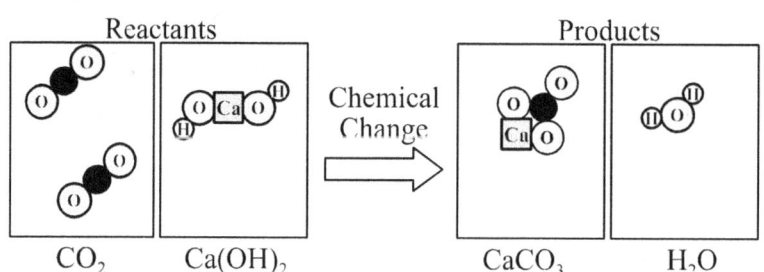

Reactants → Chemical Change → Products

CO_2 $Ca(OH)_2$ $CaCO_3$ H_2O

Balance the picture, so that it tells the whole story...

Chemical Behaviour

Science and Pedagogy

The Learning Activity Before the experiments, the student will:

Predict: Will the carbon dioxide produced by your breath, combustion of charcoal, and the reaction of vinegar and baking soda all be the same?
Explain: Use the Dalton's theory of chemical change to give a reason for your prediction.

The test for carbon dioxide: Prepare a test tube containing 10 mL of limewater (saturated solution of calcium hydroxide)

Introduce some of the gas to be tested into the test tube. Seal the tube, and shake it. If a white, cloudy precipitate of calcium carbonate forms, then the gas is likely rich in carbon dioxide.

Then the student will:
1. Place a heap of sodium bicarbonate about the size of a fingernail into a sandwich bag.
2. Put 5 mL of vinegar into a disposable plastic pipette, and without spilling a drop, put the pipette into the bag, and seal the bag.
3. Squeeze the pipette, allowing the acetic acid to mix with the sodium bicarbonate. The bag inflates at this point.
4. Test the gas by squeezing some of the gas into a test tube containing limewater.
5. Test his/her breath by gently blowing bubbles through some fresh limewater with a straw.
6. Test the gas produced when charcoal burns. A lump of glowing charcoal held in a gas bottle for a minute or so should suffice.

After the experiment
Observe and record observations, including both a description of the reaction itself, and the results of the test for carbon dioxide.
Explain observations, using the both a particle diagram and sentences.

Equipment, Preparation and Resources

12 gas bottles	12 deflagrating spoons
12 disposable pipettes	12 sandwich bags
12 drinking straws	36 test tubes
vinegar (5 % acetic acid)	sodium bicarbonate
limewater, 500 mL	

Categories: | **Assessment and Evaluation**
Knowledge: | explanation box 4, or answers to Questions for Later
Inquiry: |
Communication: | quality of writing, diagrams in explanation
Applications, Extensions: |

© Ross Lattner Publishing   www.rosslattner.ca

Explaining Chemical Change

9 Academic Science Teachers' Guide

Lab 3.9: Does Mass Change During a Chemical Reaction?

Try this pre-lab activity:

Have 7 different students measure the mass of the same sealed flask of water, on the same balance, on your lab bench.

Keep a "secret" list of the measurements, and post it only when all of the measurements have been made.

The actual mass of the water is obviously the same in every case. Why is there variation in the measured mass? How much variation is there? In which digit?

How close do two measurements have to be, in order to be "the same?"

What happens when you average the seven measurements? Do you really have 12 significant figures?

Learning Expectations: CH1.02 Student applies basic premises of the particle theory of matter to the conservation of mass problem.

Pedagogical Issues

Every chemical change produces a new substance, new kinds of particles. If there is a new substance produced, would there not be a new mass as well? It is surprising how much this issue confounds students. Its opposite misconception is equally common: When a spoonful of sugar dissolves in a glass of water, many students will predict that the mass will decrease.

When asked to give an example of matter, students will overwhelmingly choose a solid. Solids hold a special, archetypal place in children's everyday conceptualization of the world. When a solid appears or disappears, then, a great many students will infer that matter has appeared or disappeared, and that the mass has changed.

This misconception, being a kind of hybrid of science classroom thinking (matter has mass) and everyday thinking (matter is solid), is quite convincing to students. It is the purpose of this lab to challenge that idea directly, by moving the discussion into the realm of the particle notation of Dalton's theory.

Science Issues

No two measurements are exactly the same. How much alike must measurements be in order to be considered "the same"?

The point here is that students may mistake a normal variation in measurement for "proof that mass changes," a result completely at odds with the intended point.

© Ross Lattner Publishing   www.rosslattner.ca

Chemical Behaviour

Science and Pedagogy

The Learning Activity Before the experiment, the student will:

Predict: Will the mass of the flask increase, decrease, or remain the same during the chemical change?
Explain: Use the Dalton's theory of chemical change in diagrams, and sentences.

In the experiment, the student will:

1. Seal up a flask as shown, containing two different reactants.
2. Measure its mass as accurately as possible.
3. Mix the chemicals by tipping the flask.
4. Measure the mass again, to see if any new substances formed added or removed mass from the flask.

After the experiment
Observe: the masses before and after, and record.
Explain: the results in using pictures and sentences.

Equipment, Preparation and Resources

A triple beam balance is accurate enough for this lab, and with a capacity for 610 g, it can handle the combined mass of the test tube, the stopper, the 250 mL flask, and the 50 mL of solutions

12 triple beam balances   12 250 mL Erlenmeyer flasks
12 solid stoppers to fit   12 small test tubes

Some solution combinations to try:

300 mL 0.1 M $Pb(NO_3)_2$ 300 mL 0.2 M NaI
300 mL 0.1 M CaCl2   300 mL 0.1 M Na_2CO_3
300 mL 0.1 M $CuSO_4$ 300 mL 0.1 M NaOH

The students need to plan what they are doing. Many kids won't notice that they need to use the same balance, for example, or that even a small leak or spill ruins the results.

Categories:	Assessment and Evaluation
Knowledge:	explanation, box 4, and answers to Questions for Later
Inquiry:	
Communication:	quality of responses in explanation, answers to questions
Applications, Extensions:	

© Ross Lattner Publishing   www.rosslattner.ca

Student Exercises

2 Explaining Chemical Change

Knowledge and Understanding

Three theories are emphasized in this unit. Every exercise requires you to use the particle theory, Dalton's theory of chemical change, or the Bohr - Rutherford theory of the atom to solve a problem. Additional concepts will be introduced as needed.

You will also learn to use pictures to represent things that you cannot see. Your pictures, even though they may not be perfectly "true," can help you think clearly about the things you can see in your experiments. In particular, you will learn how do draw diagrams of particles, of atoms, and molecules.

Knowledge and understanding are probed at regular intervals in the Grade Nine Daily quizzes. Study these as you go through the exercises, so that you can do your best when they are assigned.

Inquiry and Thinking

We will use the PEOE cycle for most labs and activities. You are expected to frame a question, provide your best prediction, and explain your thinking, using both sentences and diagrams.

At the end of the unit, you will be given a five day independent project. The project will demonstrate your ability to conduct your own investigation.

Communication

The quality of your arguments is the most important aspect of communication in this chapter. Your arguments consist of sentences, organized into paragraphs, and supported by diagrams or other representations.

Each sentence should be clear and to the point. You will find it best to limit your sentences to two concepts linked together to make a reasonable claim. If you need to relate more than two concepts, add a new sentence.

Applications, Connections and Extensions

Every exercise in this book is designed to support you as you learn appropriate theories and apply them to problems. In the labs, you demonstrate your understanding of a theory only by applying the theory. In the quizzes and projects, you are invited to make further connections and extensions of your learning.

Explaining Chemical Change

9 Academic Science Lab Manual

Introduction: Three Theories of Chemical Change

In this unit, we will explore how one kind of matter can change into another kind of matter. Such changes, which occur all around us, are called chemical changes. This unit consists of three main ideas:

1. The **Particle Theory of Matter** consists of six simple statements which can help you explain things that happen all around you, and even help you to predict things you have never seen!

    1. **The absence of matter is a pure vacuum.** Particles of matter exist in a vacuum. The space between any two particles is always a perfect vacuum.

    2. **All matter is made of tiny particles.** Anything that has mass, and occupies space, is made of tiny particles.

    3. **All particles of one substance are identical.** Wherever you find aluminum, in foil, in airplanes, in cans, pens, anywhere - the particles of aluminum are identical to each other. Likewise, particles of corn sugar (glucose) are identical wherever you find them. Finally, the particles of glucose and aluminum must be different from each other. Surprisingly, the a single particle of glucose or aluminum may not resemble the bulk material at all.

    4. **The spaces between particles are small in solids and in liquids, and large in gases.** Particles are packed tightly in solids, less so in liquids, and are quite free to move about in gases.

    5. **All particles are attracted to each other by forces.** These forces can be weak (like the forces between two air particles) or very strong (like the forces between two iron particles).

    6. **Particles are in constant motion.** At room temperature, air molecules are crashing into each other at an average speed of 450 m/s. That's faster than a rifle bullet! Everything they crash into is set into motion with a similar amount of energy.

    The particle theory of matter is used by more people in more science-related professions than any other theory. You have probably already learned about it in previous grades.

2. **Dalton's Theory of Chemical Change** consists of six simple statements which can help you explain things that happen all around you, and even help you to predict things you have never seen! Consider the following propositions of Dalton's theory of chemical change:

    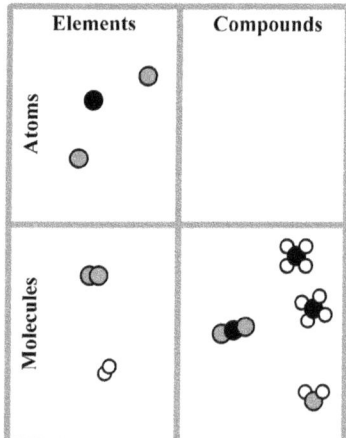

    1. An element consists of only one kind of atom.

    2. The smallest particle of a pure element is an atom.

    3. Atoms cannot be created or destroyed in chemical reactions.

© Ross Lattner Publishing   www.rosslattner.ca

Introduction

Name:
Date:

4. Compounds consist of two or more different kinds of atoms.

5. The smallest particle of a pure compound is a molecule.

6. In a chemical change, the atoms are rearranged to form new kinds out molecules.

This theory is strongly related to the particle theory. Note that Dalton's theory divides the general idea of a "particle" into different kinds of particles: atoms and molecules. With this innovation, Dalton was able to explain a great many more events and objects in the world that the particle theory alone could not explain.

While the basic propositions of Dalton's theory are still generally accepted, most of them have been developed even further in the past century. The Bohr-Rutherford theory of the atom is one of those developments. Bohr and Rutherford were the first to provide a picture of the interior of the atom which could help explain how atoms underwent chemical changes.

3. **The Bohr-Rutherford Model of the Atom.** As you have learned, atoms can combine in many different ways to make new substances. How do they do it? What's inside an atom that allows it to join with other atoms? What makes one atom different from another? As scientists, you will have to become familiar with the terms *nucleus*, *electron*, and *energy level*. You will also read about some famous scientists and their ideas of the atom. Ernest Rutherford and Neils Bohr are two scientists whose ideas are nearly 100 years old, but are still important today. The following statements outline the components of the Bohr-Rutherford model of atoms.

1. Electrons are negative, have tiny mass, and are found around the nucleus.
Protons are positive, with relatively large mass, and are found inside the nucleus.
Neutrons are neutral, with relatively large mass, and are found inside the nucleus.

2. The nucleus contains all of the positive charge (protons), and almost all of the mass (protons and neutrons) of the atom.

3. Electrons orbit the nucleus at fixed energy levels.

4. Each energy level can only hold up to 8 electrons (transition metals are exceptions).

5. Only the outermost electrons are involved in chemical change.

In all of the exercises in this book, the question must be answered in *complete sentences*. One sentence is one thought. A single word is simply not enough.

© Ross Lattner Publishing   www.rosslattner.ca

9 Academic Science
Student Exercises

Explaining Chemical Change

Lab 1.1: Particles and Physical Change

Do you Remember? List the 6 propositions of the particle theory.

1. _____
2. _____
3. _____
4. _____
5. _____
6. _____

What's The Question?
What actually happens to a substance when it melts or solidifies? We are going to observe closely as para-dichlorobenzene undergoes both melting and solidification. We won't be able to *see* heat, but we will try to pay attention to where the heat is going during the change. In the same way, we won't be able to *see* particles, but we can try to imagine what they are doing inside the test tube.

What Are We Doing?
Examine the solid menthol in the test tube provided.
1. Place one test tube containing menthol into a beaker of hot water until the menthol melts.
2. Carefully pick up the test tube of liquid menthol and hold it with your fingers until solid.

Repeat 1 and 2 as often as time permits.

What Are We Thinking About?
What is happening to the speed of the particles?

What is happening to the spaces between the particles?

What is happening to the forces between the particles?

Questions For Later... Answer all of these on a separate page.

1. Make a set of particle diagrams to show what was happening in the two changes. Label and comment on the speed, the spaces, and the forces between the particles.

2. Name the changes of state that you observed in experiments 1 and 2.

3. Was heat absorbed or released by the menthol in experiment 1? Explain your thinking.

4. Was heat absorbed or released by the menthol in experiment 2? Explain your thinking.

The Composition of Matter

Name:
Date:

Focus Question: Write the question that you are trying to answer.

1 *Predict:* Predict what will happen when you gently heat some solid menthol in a test tube.

2 *Explain* your prediction. Use the Particle Theory, in both diagrams and a paragraph.

3 *Observe,* and record your observations here.

4 *Explain your observations.* Use the Particle Theory, in both diagrams and a paragraph.

© Ross Lattner Publishing   49   www.rosslattner.ca

Explaining Chemical Change

9 Academic Science
Student Exercises

Lab 1.2: Sublimation of Iodine

Do you Remember? List the 6 propositions of the Particle Theory
1. _____
2. _____
3. _____
4. _____
5. _____
6. _____

What's The Question?
What will an iodine crystal do when heated? Will it change colour or odour? Will it turn to a liquid, remain solid, or do something else? Whatever it does, we can learn something about iodine particles.

What Are We Doing? Put a few crystals of iodine into a test tube. Insert a cork, not rubber, stopper LOOSELY. Put the test tube into some boiling water.
Predict the outcome.
Explain your prediction.
Observe the outcome.
Explain your observation.

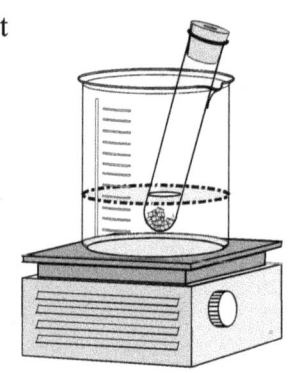

What Are We Thinking About?
1. The melting point of iodine is 113°C, and the boiling point of iodine is 184°C.
2. How are the particles of the solid iodine arranged?
3. What is happening to particles of iodine as the iodine is heated?
4. Think about the *speed* of the iodine particles, the *distance* between particles, and the *forces* between the particles.

Questions For Later...
1. Is the iodine melting, boiling, or doing something else? Explain your answer.

2. Name the two changes of state that you witnessed in this experiment?

3. At what time in this experiment were the iodine particles farthest apart from each other and closest to each other?

4. At what time in this experiment were the forces between the iodine particles weakest and strongest?

The Composition of Matter

Name:
Date:

Focus Question: Write the question that you are trying to answer.

1 Predict: Predict what will happen when you gently heat a few iodine crystals in a test tube.

2 Explain your prediction. Use the Particle Theory, in both diagrams and a paragraph.

3 Observe, and record your observations here.

4 Explain your observations. Use the Particle Theory, in both diagrams and a paragraph.

© Ross Lattner Publishing   www.rosslattner.ca

9 Academic Science
Student Exercises

Explaining Chemical Change

Lab 1.3: Classifying Elements as Metals and Non-metals

What's The Question?
What is an element? What kinds of elements are there?

Elements are the simplest forms of matter and appear to fall into two groups. *Metals* are shiny, silvery, malleable, and good conductors of heat and electricity. *Non-metals* are dull, have various colours, are brittle, and are poor conductors of heat and electricity.

What Are We Doing?
You will be given a collection of 12 elements.

You will examine them, and record their properties. On the basis of your observations, you will classify these elements as metals or non-metals.

What Are We Thinking About?
1. A *mallet* is a hammer.

2. *Malleable* materials can be hammered without breaking. They dent or bend.

3. *Non - malleable* materials cannot be hammered. They break or shatter.

Questions For Later...

1. Using your table as a guide, list the names of all elements that have *all three* of the following properties: **shiny, malleable** and **able to conduct electricity**.

2. Are the substances in Question 1 *metals*, or *non-metals*? Explain.

3. Using your table as a guide, list the name of all elements that have *all three* of the following properties: **dull, not malleable** and **non-conducting**.

4. Are the substances in Question 3 *metals*, or *non-metals*? Explain

The Composition of Matter

Name:
Date:

Element Name	Lustre *Shiny* or *Dull*	Malleable *Dents* or *Shatters*	*Conductor* or *Non-conductor*
1			
2			
3			
4			
5			
6			
7			
8			
9			
10			
11			
12			

All of the *elements* above consist of just *one kind of atom*.

For Our Next Performance...
Matter can be classified into two main groups: those things which are *pure*, and those which are *mixtures*.

Pure substances contain only one kind of particle. They are either elements (1 kind of **atom**) or compounds (1 kind of **molecule**).

Mixtures contain two or more kinds of particles. In **solutions** the particles of the substances are thoroughly mixed, and no chunks larger than molecules exist. In all **other mixtures** particles are still in large clumps, and the clumps are mixed together.

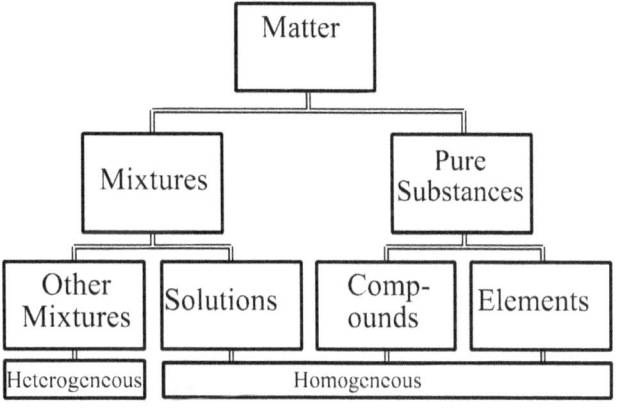

Homogeneous substances always have the same composition throughout. Whether they are pure elements, pure compounds, or solutions, they are composed of the same kind of matter throughout.

Heterogeneous substances have variable composition. A bit taken from one part of the mixture might have different composition from another bit. Heterogeneous substances often have lumps or chunks of various sizes. These mixtures scatter light, appear cloudy and often have visible chunks in them.

© Ross Lattner Publishing

9 Academic Science Student Exercises

Explaining Chemical Change

Lab 1.4: Classifying Substances by Composition

What's The Question?
Almost every substance that you can imagine can be classified into one of the four major categories *element, compound, solution, mixture*. How do we scientists tell these kinds of matter apart?

What Are We Doing?

Recipe for Play Dough
250 mL water   125 mL fine salt
2 tbsp vegetable oil
2 tbsp alum   600 mL white flour
food colouring (red, green, yellow, and blue) or
dry tempera paint (red, green, yellow, blue, black)

1. Heat the water to boiling in a stainless pot.
2. Add salt; heat to boiling.. Add alum.
3. Add vegetable oil Stir.
4. Remove from heat and immediately mix in flour, a little at a time, until it is all stirred in.
5. Turn out onto a floured board. Let cool. Knead until smooth.
6. Add colouring and knead in. For more intense colour, use tempera paint.
7. Store in air tight containers. Dries to make a hard solid material.

What Are We Thinking About?
Composition means "What makes up a substance".

An atom is the smallest particle of an element. We can use marble-sized balls of play dough to represent these elements:

Sulfur   yellow   Carbon   black
Oxygen   red   Chlorine   green
Metal   blue   Hydrogen   white

What is a chemical formula? What does it tell you about the kinds of atoms in a substance?

Complete the classification tree diagram below by adding the words *heterogeneous, homogeneous, pure substance, mixture, element, solution, compound, matter,* and *mixture*.

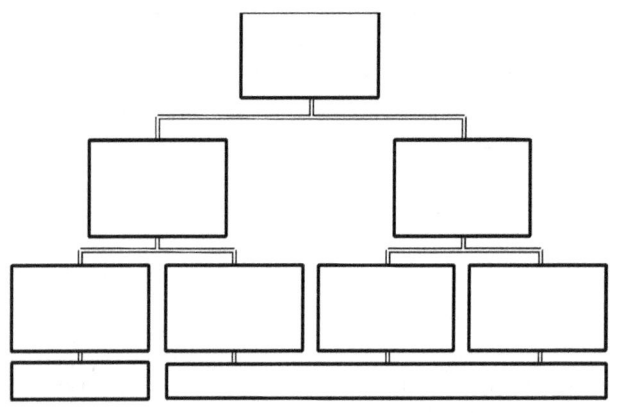

Questions for Now...

1. What is the difference between an *atom*, and a *molecule*?

2. How many different kinds of atoms are there in an *element*?

3. How many different kinds of atoms are there in a *compound*?

4. How many different kinds of molecules are there in a pure *compound*?

The Composition of Matter

Name:
Date:

Substance	Chemical Formula	Element, Compound, Solution, or Mixture? Reasons for your choice
Water	H_2O	
Table Salt	NaCl	
Salt water		
Vegetable Oil	$C_{12}H_{23}COOH$	
Alum	$KAl(SO_4)_2$	
Flour		
Red Food Colour		
Play Dough		

5. What is the main characteristic that you used to distinguish the *mixtures* from *solutions*?

6. Make a full page table like the one below. Your teacher will supply you with a list of common substances, and their chemical formulas. Make a play dough model of each substance. Then write the name, formula, and picture of each substance in one of the four boxes.

	Elements	Compounds
Atoms	Both an *Atom* and an *Element*	Both an *Atom* and a *Compound*
Molecules	Both a *Molecule* and an *Element*	Both a *Molecule* and a *Compound* Water, H_2O, ⚬⚬

© Ross Lattner Publishing   www.rosslattner.ca

9 Academic Science
Student Exercises

Explaining Chemical Change

Lab 1.5: The Electrolysis of Water

Do you Remember? Define the following terms:

1. Element _____
2. Compound _____
3. Atom _____
4. Molecule _____

What's The Question?
All around you substances are changing into other substances. Charcoal in a barbecue changes into ashes; vegetable scraps change into compost; iron changes into rust. What is actually happening when one kind of substance changes into another kind of substance?

What Are We Doing?
Two test tubes are filled with water and inverted over the electrodes. A current is passed through as indicated. **Caution: some sodium hydroxide is added to the water to improve conductivity.**

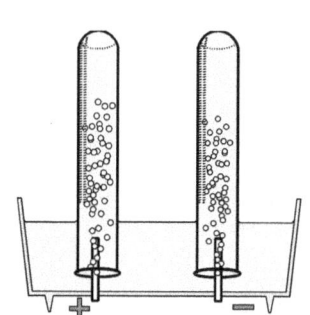

What Are We Thinking About?
Electrolysis is the decomposition of a compound by passing an electric current through it.

1. Is water an element or a compound?

2. Is the smallest particle of water an atom, or a molecule?

Questions For Later...

1. Which generated gas had the biggest volume, hydrogen or oxygen?

2. What is the ratio of volumes of hydrogen to oxygen?

3. What is the chemical formula for water?

4. Does the ratio of the volumes of hydrogen and oxygen make sense once you look at the chemical formula? Explain.

The Composition of Matter

Name:
Date:

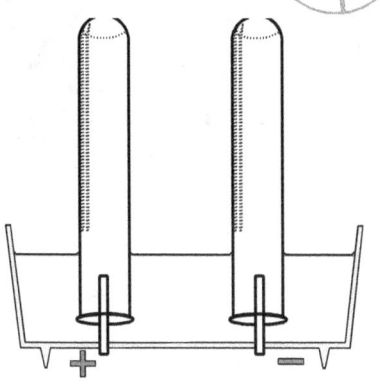

What do you see?
1. Complete the diagram at right to show the volumes of the gases in each test tube.
2. How many times bigger was the larger volume than the smaller one? (3 times as big? 4 times as big?)
3. If you increase the current flow, do you get a different ratio of gases?

What Gases Are Being Produced?
1. Perform the test for oxygen gas on both test tubes. Describe your observations below.
2. Perform the test for hydrogen gas on both test tubes. Describe your observations below.
3. Label the gases in the diagram above.

Consider the following propositions of Dalton's Theory of Chemical Change:
1. **An element consists of only one kind of atom.**
2. **The smallest particle of a pure element is an atom.**
3. **Atoms cannot be created or destroyed in chemical reactions.**
4. **A compound consists of two or more different kinds of atoms.**
5. **The smallest particle of a pure compound is a molecule.**
6. **In a chemical change, the atoms are rearranged to form new kinds of molecules.**

Complete the diagram at right, to show how the water molecules have been changed into the elements oxygen and hydrogen. Be sure to count up all of your atoms! How many hydrogen molecules, and how many oxygen molecules, can be made?

Whenever we use this kind of diagram, you must complete it in the following way:
1. Mark each box either *Element* or *Compound* (*E* or *C*)
2. In addition, mark each box either *Atoms* or *Molecules* (*A* or *M*)
3. Below the arrow, write some words to describe the visible signs of a chemical change.

© Ross Lattner Publishing 57 www.rosslattner.ca

All the news that's fit to print... and then some

The Grade Nine Daily

Quiz 1.6: Dalton's Theory

Some of the answers for this page can be found in the list at the bottom of the page.

1 Complete the Classification of Matter chart

Date: _____ / 4

2 Classify this system as **E**lement, **C**ompound or **M**ixture. Shade in one box in each row.

Circle two different kinds of molecules. Are they **E**lements or **C**ompounds?
Could this system be a **S**olution? Be prepared to explain.

Date: _____ / 4

3 In the list below are things that you might find in a restaurant. Are they **E**lements **C**ompounds, **S**olutions or **M**echanical mix?

1. ____ salt
2. ____ vinegar
3. ____ coffee
4. ____ milk
5. ____ pepper
6. ____ ketchup
7. ____ relish
8. ____ plastic spoon
9. ____ sugar
10. ____ silver spoon

Date: _____ / 4

4 Classify each box as containing **E**lement, **C**ompound or **M**ixture

Date: _____ / 4

Element	Atom	Solution	Heterogeneous
Compound	Molecule	Pure Substance	Homogeneous
Suspension	Mixture	Mechanical Mixture	Colloid

© Ross Lattner Publishing   www.rosslattner.ca

The Grade Nine Daily

All the news that's fit to print... and then some

Quiz 1.6: Dalton's Theory Name:

---

5 List 5 visible signs of a chemical change:

1. _____
   _____
2. _____
   _____
3. _____
   _____
4. _____
   _____
5. _____
   _____

Date: _____ / 4

---

6 Is it a **P**hysical or **C**hemical change?

1. _____ an ice cube melts
2. _____ a light bulb lights up
3. _____ a match burns
4. _____ a car rusts
5. _____ frost forms on a window
6. _____ a log rots in a forest
7. _____ a plant grows
8. _____ milk sours in the fridge
9. _____ a drinking glass is broken
10. ____ a cake bakes

Date: _____ / 4

---

7 Mark arrows **P**hysical or **C**hemical change.

Date: _____ / 4

---

8 Connect the **R**eactant box on the left with the **P**roduct box on the right.

Date: _____ / 4

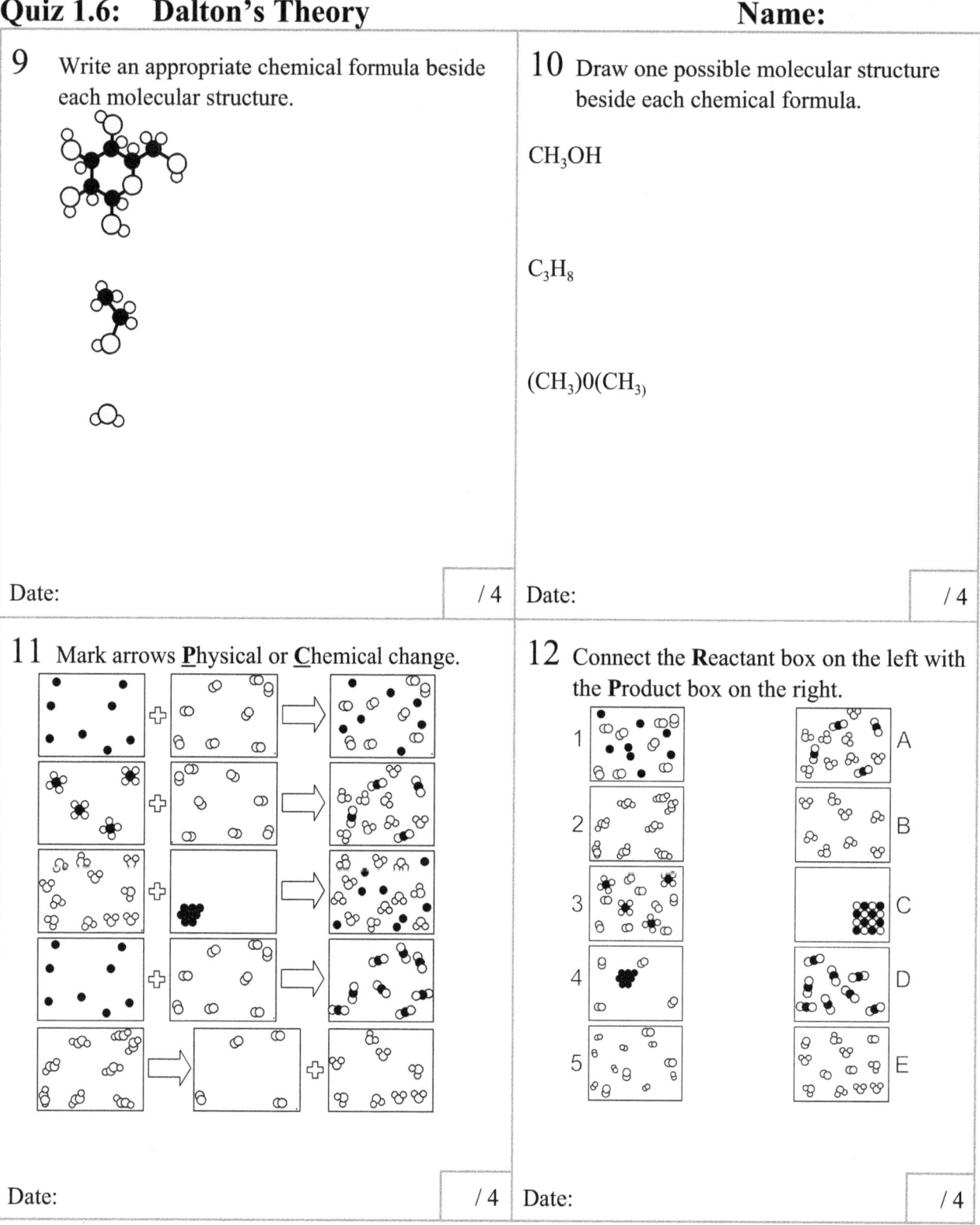

All the news that's fit to print... and then some

The Grade Nine Daily

Quiz 1.6: Dalton's Theory Name:

13 Classify this system. Fill in one box in each row to indicate that classification.

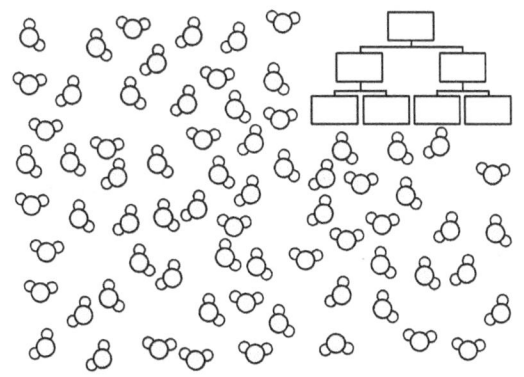

How many different kinds of molecules are there? Circle two different kinds of atoms.

Date: _____ / 4

14 Classify this system. Fill in one box in each row to indicate that classification.

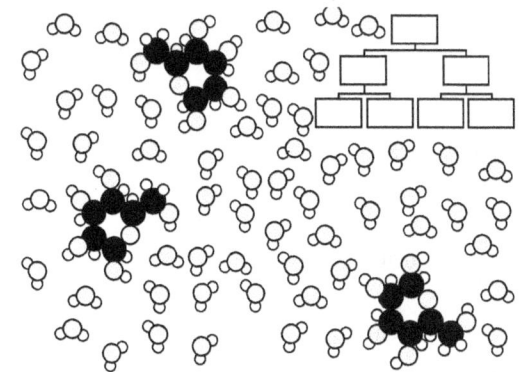

Circle two different kinds of molecules.

Date: _____ / 4

15 Connect **R**eactants left with **P**roducts right.

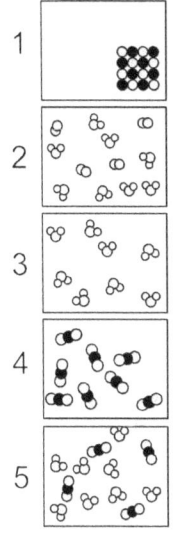 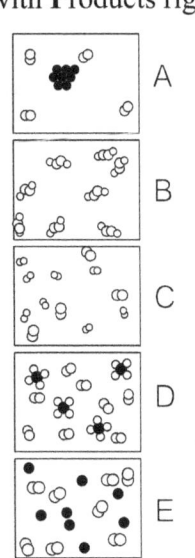

Date: _____ / 4

16 Mark arrows **P**hysical or **C**hemical change.

Date: _____ / 4

© Ross Lattner Publishing   www.rosslattner.ca

Explaining Chemical Change

9 Academic Science Student Exercises

Lab 2.1: The Bohr-Rutherford Model of the Atom

Do you Remember? List the 6 propositions of Dalton's Theory of Chemical Change.

1. _____
2. _____
3. _____
4. _____
5. _____
6. _____

What's The Question?

We know that from only 92 naturally occurring elements, millions of compounds can form. Each of the 92 elements is unique. In this experiment, we will heat atoms of elements in very hot flames, and observe their characteristic colours. Each of the 92 elements has its own set of colours.

What are atoms made of? What are they like inside? Whatever model we scientists think up to answer these questions must also explain everything else we know about matter.

What Are We Doing?

1. Your teacher will either set up Bunsen burners, or tell you how to safely light your own.
2. Choose a wooden splint that has been soaked in a compound. *Do not mix chemicals!!*
3. Hold the wooden splint in the Bunsen burner flame for a few seconds. Observe and record the colour.
4. When the stick begins to burn, plunge it back into the proper container to wet it again.
5. Repeat with other chemicals, until you have tested all of the solutions.

What Are We Thinking About?

1. What are the positive particles in the atom?
2. Where are the positive particles found?
3. Where are the negative particles found in the atom? What are they called?
4. Describe the force between the electrons and the protons in the atom.

Questions For Later...

1. Do all compounds containing copper always make the same colours?

2. You would like to test a small sample of soil to see if it contained strontium. What would you do with the soil? How would you know that it contained strontium?

3. Why do different elements have different colours in the flame? Use the words *proton*, *electron*, *neutron*, and *energy level* in your answer.

The Periodic Table

Name:
Date:

Focus Question: Write the question that you are trying to answer.

1 Experiment: Test each of the solutions provided by your teacher. Carefully record the name of the metal in the compound, and the colour of the compound.

Sample	Metal	Description of colour
1		
2		
3		
4		
5		
6		
7		
8		
9		
10		
11		
12		

2 Explain: Why does each atom have its own unique colour? Draw pictures to support your answer.

3 Refine your Explanation using diagrams and sentences.

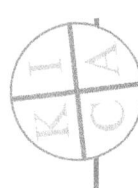

Name:
Date:

The Periodic Table

Activity 2.2: Organization of the Periodic Table

Look up the term periodic table or periodic table of the elements in the index of your text book or another reference. Use your text to complete the periodic table on the opposite page.

1. In each rectangle, print neatly all of the following information: *Name*, *Symbol (X)*, *Atomic Number (Z)*, *Mass Number (A)*.

2. Label each column 1, 2, 3, ... 18 etc, according to the table in your text book.

3. Obtain three coloured markers, or highlighters. Highlight all of the *non-metals* one colour, the *metals* another colour, and the borderline cases, the *metalloids*, a third colour. What pattern do you notice?

4. A large group of elements are called the *transition metals*. Where are they on the periodic table? Look them up in your text book, in an encyclopedia, or the Internet.

5. Label the *alkali metal* group or family of elements.

6. Label the *alkali earth metals* group or family of elements.

7. Label the *halogens* and the *noble gases*.

$^{Z}_{A}X$ Name

$^{7}_{14.01}N$ Nitrogen

The Periodic Table

Name:
Date:

Activity 2.3: Bohr, Rutherford and the Periodic Table

Do you Remember? What are the parts of the Bohr-Rutherford model of the atom? Where is each part found within the atom?

1.
2.
3.

What's The Question?

The Bohr-Rutherford model tells us a great deal about the structure of individual atoms. *What is the relationship between the Bohr-Rutherford model of the atom, and the structure of the periodic table?*

What Are We Doing?

1. Fill in the electrons as shown in the example. The first two electrons go to the innermost level, where they are closest to the attractive positive nucleus. Additional electrons go into the next closest level until it is filled, and so on.
2. Print the number of protons (+) and neutrons (n) inside the nucleus.
3. Print the symbol and the name for the element.
4. Label the Group at the top of each column (1, 2, 13, 14, ...etc.)
5. Label the Row at the left of each period (1, 2, 3, ... etc.).

What Are We Thinking About?

Both protons and neutrons have almost exactly the same mass. Protons and neutrons are relatively massive, about 2000 times more massive than electrons. Protons carry one positive charge, while neutrons carry no charge at all.

According to the Bohr-Rutherford model, the nucleus is very tiny. In the picture at left, it should be about one-one-thousandth the size shown, but that would be too small see in the diagram. Despite its tiny size, it contains all of the positive charge, and nearly all of the mass of the atom. The electrons carry all of the negative charge, and a very tiny amount of mass.

Questions For Later...

1. What is the same about the electron arrangement of every element in a column (group)? What is different?

2. What is the same about the electron arrangement of every element in a row (period)? What is different?

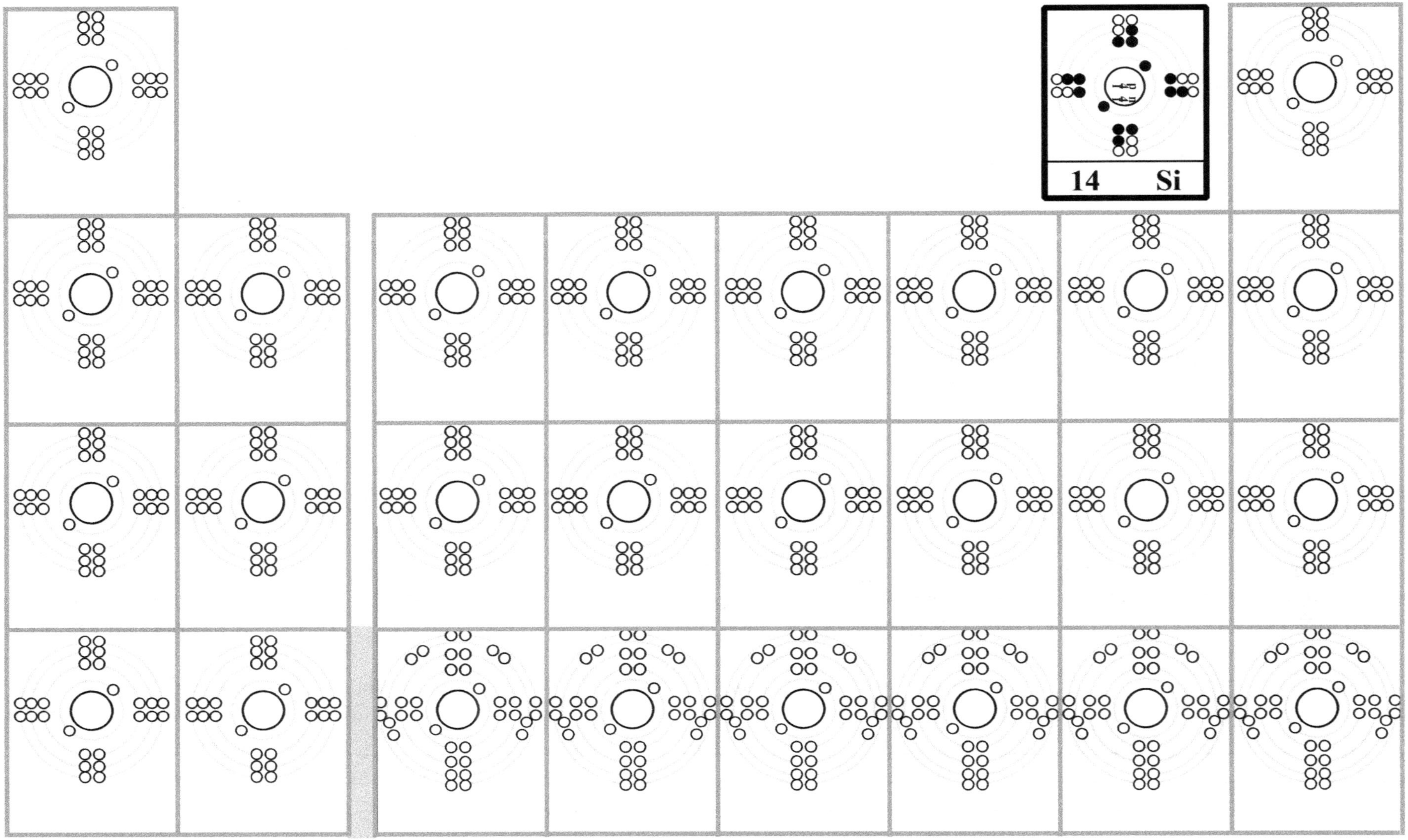
14
Si

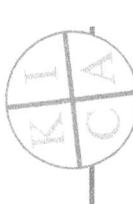

Name:
Date:

The Periodic Table

Activity 2.4: Valence Electrons, Atomic Cores and the Periodic Table

Do you Remember? At the beginning of the unit on electricity, we studied a theory of the electrical structure of matter. Review it, and write it here.

_____

_____

What's The Question?
The Bohr-Rutherford model of the atom has a lot of parts. *Can we simplify the model to make it easier to use?*

What Are We Doing?
1. Fill in the *valence* electrons as shown in the example. The valence electron in the example is the electron in the outermost, partially filled shell.
2. Print the charge on the *atomic core*. This is the sum of all of the positive charges in the nucleus plus all of the negative charges on the core electrons. In this case, 10 core electrons plus 11 protons gives a core charge of positive one.
3. Print the symbol and the name for the element.
4. Label the group at the top of each column (1, 2, 13, 14...etc.).
5. Label the row at the left of each period (1, 2, 3 etc.).

What Are We Thinking About?
1. Chemists have found that only the electrons in the outermost, partially filled shell directly affect the chemistry of the atom. These outermost electrons are called the *valence electrons*. All of the other electrons in the inner, completely filled shells, are called *core electrons*.
2. In the example at left, there is one valence electron and ten core electrons. The ten core electrons plus the 11 protons make up the atomic core. The overall charge on the atomic core is +11 −10, for a total charge of +1.

Questions For Later...
1. What is the relationship between group number, and the number of valence electrons?

2. What is the relationship between group number, and the charge on the atomic core?

14 Si +4

The Periodic Table

Name:
Date:

Activity 2.5: Atomic Radius and the Periodic Table

What's The Question? *How does the changing atomic core charge affect the size of the atom?*

What Are We Doing?
1. Plot the radii from the table onto the graph.
2. Colour the metals blue, the non-metals red.

#	Symbol	Atomic Radius
1	H	0.032
2	He	0.031
3	Li	0.152
4	Be	0.089
5	B	0.082
6	C	0.077
7	N	0.075
8	O	0.073
9	F	0.072
10	Ne	0.071

#	Symbol	Atomic Radius
11	Na	0.186
12	Mg	0.136
13	Al	0.118
14	Si	0.111
15	P	0.106
16	S	0.102
17	Cl	0.099
18	Ar	0.098
19	K	0.227
20	Ca	0.174

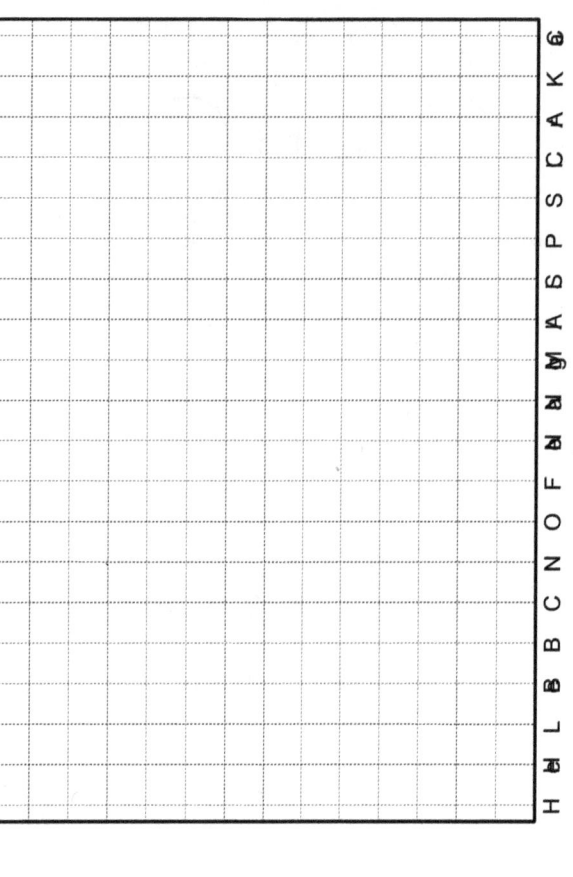

3. In the periodic table below, draw a circle to represent the size of the atom. Let 1 mm represent 0.01 nm. In that scale, Silicon, with radius 0.111 nm, would be drawn as a circle with a radius of 11.1 mm. (See example below). Write the element symbol in each case. Colour the non-metals red, the metals blue, and the metalloids yellow.

Questions For Later...
1. How does atomic radius change as you read from left to right across a row? Why?
2. How does atomic radius change as you read down a family? Why?

| 14 Si 11.1 mm |

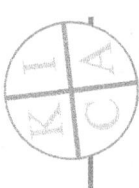

The Periodic Table

Name:
Date:

Activity 2.6: Putting it All Together

What's The Question? *Why are some atoms metals, others non-metals? How does this affect each element's behaviour?*

Metal Behaviour	*Non-Metal Behaviour*

Metal Behaviour

A metal like lithium has a small atomic core charge, and a large radius. Its valence electron is very far from a weak charge, and is therefore easy to remove.

Metals frequently give up their weakly held electrons to non-metal atoms. This gives rise to rusting, tarnishing, and corrosion. For example, when iron on your bicycle loses electrons to oxygen in the air, rust-coloured iron oxide is formed.

Because the electrons are easy to move, in a solid metal they can easily move from atom to atom. Electrons which are easy to move give rise to all of the common behaviours of metals: they are shiny, good conductors of heat and electricity, and bend rather than shatter.

Non-Metal Behaviour

A non-metal atom like fluorine has a large core charge, and a small radius. Its electrons are very tightly held. In fact, fluorine is so small and has such a strong core charge that it is able to attract and hold the electrons of other atoms.

Oxygen, for example, can grab and hold iron's electrons, to form rust. Oxygen grabs carbon's electrons in a candle flame. The oxygen you breathe can attract electrons from the sugar that you eat, and provide you with the energy that you need to live and move.

Non-metals are almost always poor conductors of heat and electricity. Apart from that, their variety is amazing. Non-metals can be solids, liquids, or gases, clear, colourless, or opaque, brittle or flexible.

In the periodic table below:
1. complete each box by drawing in the valence electrons, printing the atomic core charge, and the name, number and symbol of the element.
2. Colour the atomic cores of the metals blue, non-metals red, and metalloids yellow.
3. Electronegativity is *the ability to attract and hold electrons from another atom*. Look up the electronegativity of each element, and write it in the top right corner for each element.

Questions For Later...

1. List ten objects you can see right now that are metals, and ten that are non-metals.

2. Would you expect the chlorine found in water to be shiny, hard, flexible, and a good conductor? Explain.

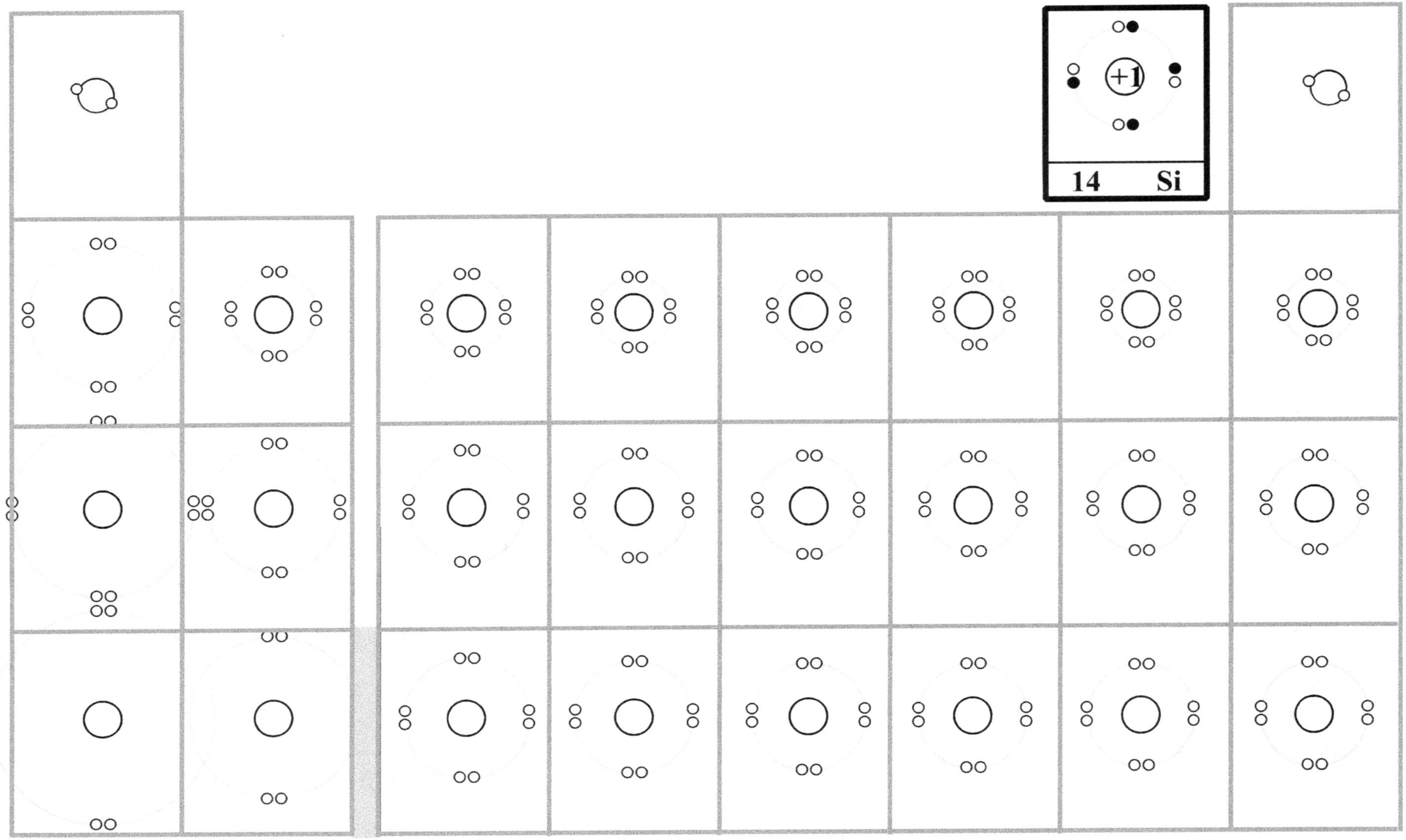
+1
14
Si

Resource 2.7: The Standard Periodic Table

Main Table

Legend:
- Atomic Number (top left)
- Electronegativity (top right)
- Symbol
- Name
- Mass

Group	1	2	3	4	5	6	7	8	9	10	11	12	13	14	15	16	17	18
	1 H 2.1 Hydrogen 1.01																	2 He – Helium 4.00
	3 Li 1.0 Lithium 6.9	4 Be 1.5 Beryllium 9.01											5 B 2.0 Boron 10.8	6 C 2.5 Carbon 12.01	7 N 3.0 Nitrogen 14.01	8 O 3.5 Oxygen 16.00	9 F 4.0 Fluorine 19.0	10 Ne – Neon 20.2
	11 Na 0.9 Sodium 23.0	12 Mg 1.2 Magnesium 24.3											13 Al 1.5 Aluminum 27.0	14 Si 1.8 Silicon 28.1	15 P 2.1 Phosphorus 31.0	16 S 2.5 Sulfur 32.1	17 Cl 3.0 Chlorine 35.5	18 Ar – Argon 39.9
	19 K 0.8 Potassium 39.1	20 Ca 1.0 Calcium 40.1	21 Sc 1.3 Scandium 45.0	22 Ti 1.5 Titanium 47.9	23 V 1.6 Vanadium 50.9	24 Cr 1.6 Chromium 52.0	25 Mn 1.5 Manganese 54.9	26 Fe 1.8 Iron 55.8	27 Co 1.8 Cobalt 58.9	28 Ni 1.8 Nickel 58.7	29 Cu 1.9 Copper 63.5	30 Zn 1.6 Zinc 65.4	31 Ga 1.6 Gallium 69.7	32 Ge 1.8 Germanium 72.6	33 As 2.0 Arsenic 74.9	34 Se 2.4 Selenium 79.0	35 Br 2.8 Bromine 79.9	36 Kr – Krypton 83.8
	37 Rb 0.8 Rubidium 85.5	38 Sr 1.0 Strontium 87.6	39 Y 1.2 Yttrium 88.9	40 Zr 1.4 Zirconium 91.2	41 Nb 1.6 Niobium 92.9	42 Mo 1.8 Molybdenum 95.9	43 Tc 1.9 Technicium (98)	44 Ru 2.2 Ruthenium 101.1	45 Rh 2.2 Rhodium 102.9	46 Pd 2.2 Palladium 106.4	47 Ag 1.9 Silver 107.9	48 Cd 1.7 Cadmium 112.4	49 In 1.7 Indium 114.8	50 Sn 1.8 Tin 118.7	51 Sb 1.9 Antimony 121.8	52 Te 2.1 Tellurium 127.6	53 I 2.5 Iodine 126.9	54 Xe – Xenon 131.3
	55 Cs 0.7 Cesium 133	56 Ba 0.9 Barium 137.3	71 Lu 1.2 Lutetium 175	72 Hf 1.3 Hafnium 178.5	73 Ta 1.5 Tantalum 180.9	74 W 1.7 Tungsten 183.8	75 Re 1.9 Rhenium 186.2	76 Os 2.2 Osmium 190.2	77 Ir 2.2 Iridium 192.2	78 Pt 2.2 Platinum 195.1	79 Au 2.4 Gold 197.0	80 Hg 1.9 Mercury 200.6	81 Tl 1.8 Thallium 204.4	82 Pb 1.8 Lead 207.2	83 Bi 1.9 Bismuth 209.0	84 Po 2.0 Polonium (209)	85 At 2.2 Astatine (210)	86 Rn – Radon (222)
	87 Fr 0.7 Francium (223)	88 Ra 0.9 Radium (226)	103 Lr Lawrencium (256)	104 Unc	105 Unp	106 Unh	107 Uns	108 Uno	109 Une									

Lanthanide Series

| 57 La 1.1 Lanthanum 138.9 | 58 Ce 1.1 Cerium 140.1 | 59 Pr 1.1 Praseodymium 140.9 | 60 Nd 1.1 Neodymium 144.2 | 61 Pm 1.1 Promethium (145) | 62 Sm 1.1 Samarium 150.4 | 63 Eu 1.1 Europium 150.2 | 64 Gd 1.1 Gadolinium 157.2 | 65 Tb 1.1 Terbium 158.9 | 66 Dy 1.1 Dysprosium 162.5 | 67 Ho 1.1 Holmium 162.5 | 68 Er 1.1 Erbium 167.3 | 69 Tm 1.1 Thulium 168.9 | 70 Yb 1.1 Ytterbium 173.0 |

Actinide Series

| 89 Ac 1.1 Actinium (227) | 90 Th 1.1 Thorium 232.0 | 91 Pa 1.5 Protactirium (231) | 92 U 1.7 Uranium 238.0 | 93 Np 1.3 Neptunium (237) | 94 Pu 1.3 Plutonium (244) | 95 Am 1.3 Americium (243) | 96 Cm 1.3 Curium (247) | 97 Bk 1.3 Berkelium (247) | 98 Cf 1.3 Californium (251) | 99 Es 1.3 Einsteinium (254) | 100 Fm 1.3 Fermium (253) | 101 Md 1.3 Mendelevium (257) | 102 No 1.3 Nobelium (255) |

Ions

Ammonium NH_4^+	Cyanide CN^-	Acetate $C_2H_3O_2^-$	Nitrite NO_2^-
Hydroxide OH^-	Cyanate OCN^-	Thiosulfate $S_2O_3^{2-}$	Iodite IO_2^-
Thiocyanate SCN^-			Iodate IO_3^-

Nitrate NO_3^-	Hypochlorite ClO^-	Permanganate MnO_4^-	Sulfite SO_3^{2-}
Chlorite ClO_2^-	Hypobromite BrO^-		Hydrogen Sulfite HSO_3^-
Chlorate ClO_3^-	Bromite BrO_2^-		Sulfate SO_4^{2-}
Perchlorate ClO_4^-	Bromate BrO_3^-		Hydrogen Sulfate HSO_4^-
	Perbromate BrO_4^-		

Carbonate CO_3^{2-}	Chromate CrO_4^{2-}	Phosphate PO_4^{3-}	
Hydrogen Carbonate HCO_3^-	Dichromate $Cr_2O_7^{2-}$	Hydrogen Phosphate HPO_4^{2-}	
		Dihydrogen Phosphate $H_2PO_4^-$	

All the news that's fit to print... and then some
The Grade Nine Daily

Quiz 2.8: Periodic Table

Some of the answers for this page can be found in the list at the bottom of the page.

1 Complete the Bohr-Rutherford diagram below for the atom aluminum-27. $^{27}_{13}Al$

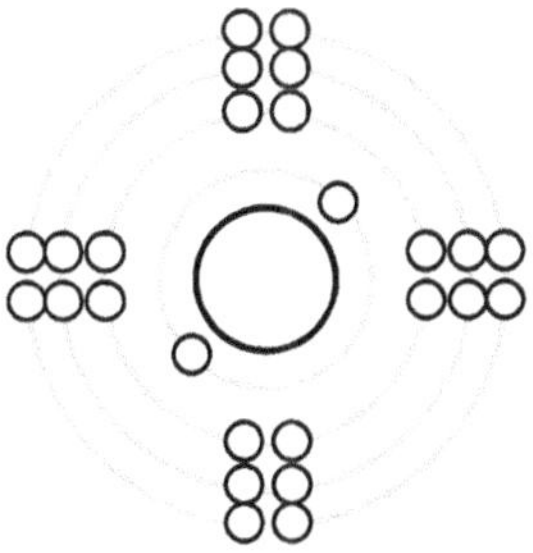

_____ Number of heavy particles in the nucleus
_____ Number of protons
_____ Number of neutrons

Date: _____ / 4

2 Complete the Bohr-Rutherford diagram below for the atom oxygen-16. $^{16}_{8}O$

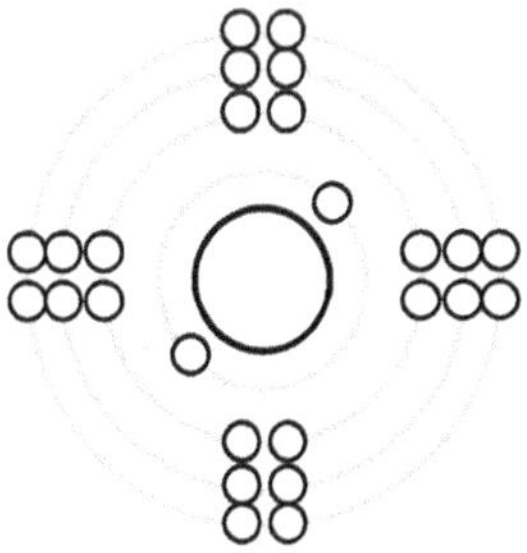

_____ Number of heavy particles in the nucleus
_____ Number of protons
_____ Number of neutrons

Date: _____ / 4

3

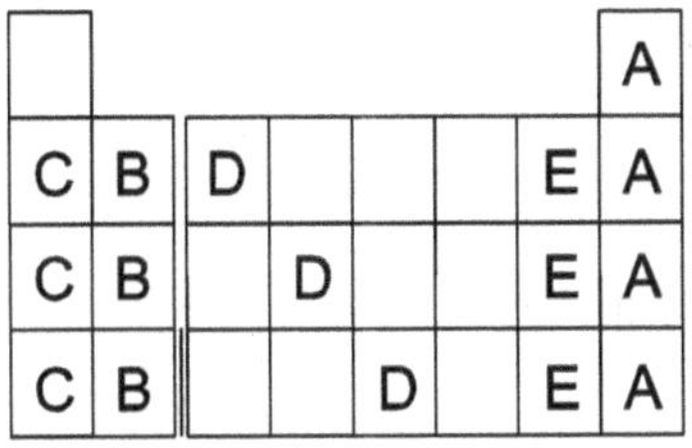

Which *family* or *group* of elements is indicated by each letter?

A _____   D _____

B _____   E _____

C _____

Date: _____ / 4

4

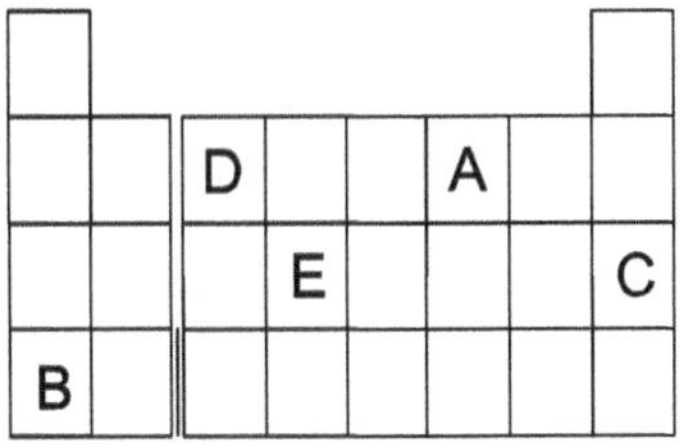

How many electrons are present in the valence shell for each of the indicated atoms?

A _____   D _____

B _____   E _____

C _____

Date: _____ / 4

Metal	Alkali metal	Noble Gas	Electron
Non-metal	Alkali earth metal	Proton	Heavy Particle
Metalloid	Halogen	Neutron	Shell

© Ross Lattner Publishing   75   www.rosslattner.ca

The Grade Nine Daily

All the news that's fit to print... and then some

Quiz 2.8: Periodic Table   Name:

You may need to refer to a simple periodic table to answer these questions.

---

5 Complete the Bohr-Rutherford diagram below for the atom potassium-39. $^{39}_{19}K$

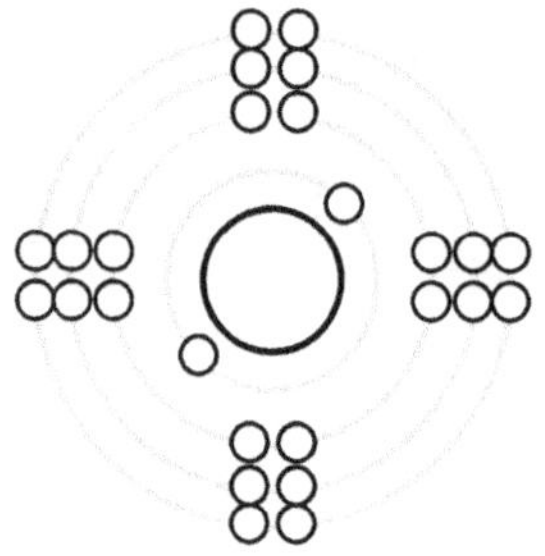

_____ heavy particles in the nucleus
_____ Number of protons
_____ Number of neutrons
_____ Number of electrons

Date: / 4

---

6 Complete the Bohr-Rutherford diagram below for the atom calcium-40. $^{40}_{20}Ca$

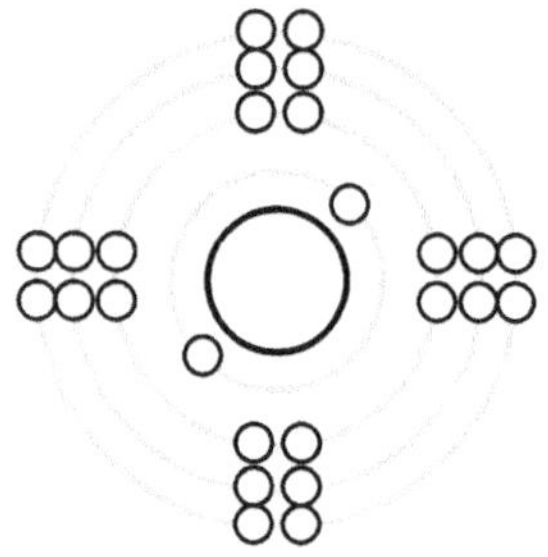

_____ heavy particles in the nucleus
_____ Number of protons
_____ Number of neutrons
_____ Number of electrons

Date: / 4

---

7 Mark one box in the table with each letter:

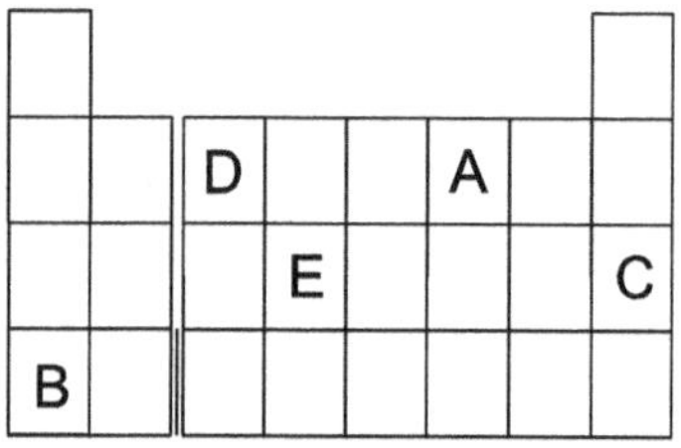

A   same row, smaller than A
B   one more valence electron than B
C   same group as C, but smaller
D   same row as D, but more like a metal
E   same group as E, but larger

Date: / 4

---

8 Mark all of the boxes with the appropriate letters to indicate:

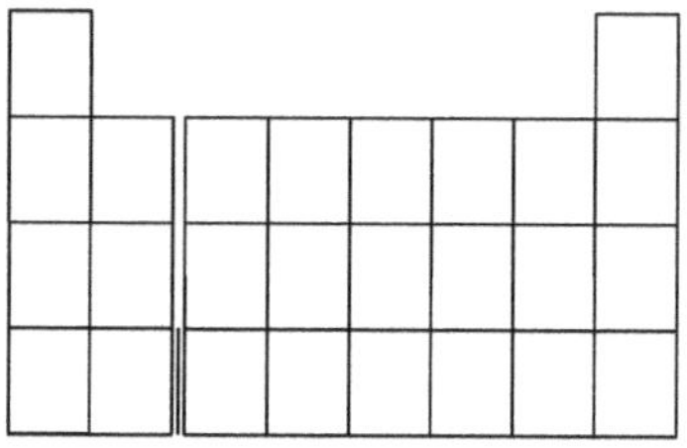

A   all of the Noble Gases
B   all of the Alkali Metals
C   all of the Halogens
D   all of the Metalloids
E   all of the Alkali Earth metals

Date: / 4

All the news that's fit to print... and then some
The Grade Nine Daily

Quiz 2.8: Periodic Table   Name:

You may need to refer to a simple periodic table to answer these questions.

---

9 Complete the valence-core diagram below for the element sulfur. $^{32}_{16}S$

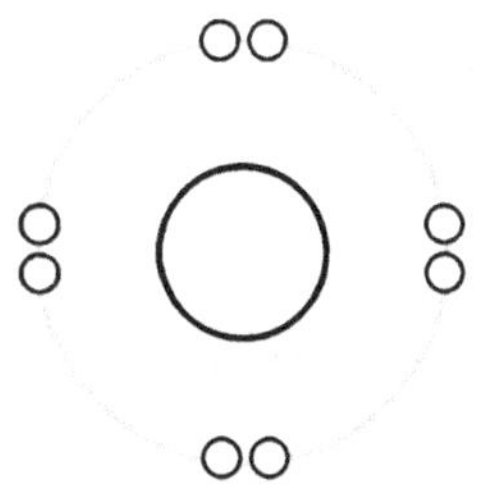

_____ positive charges in the atomic core
_____ number of valence electrons

Date: _____   / 4

---

10 Complete the valence-core diagram below for the element carbon. $^{12}_{6}C$

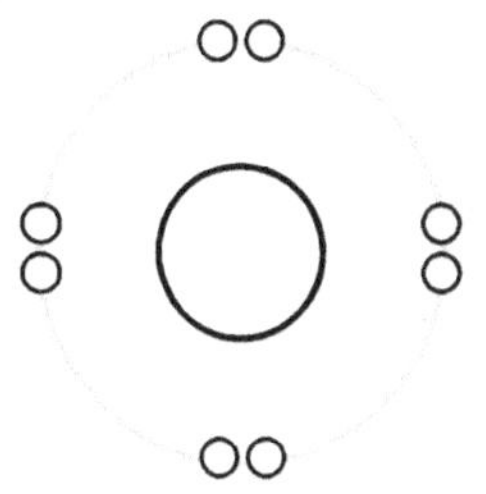

_____ positive charges in the atomic core
_____ number of valence electrons

Date: _____   / 4

11

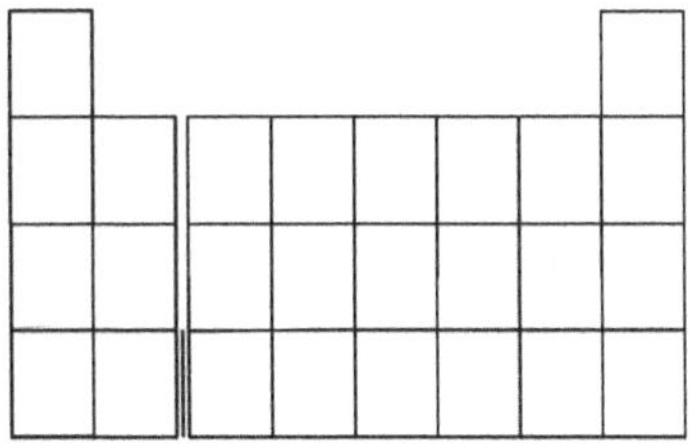

Mark boxes with appropriate letters to indicate:

A   the element with only one electron
B   the smallest Noble gas
C   most likely to grab and hold an extra electron
D   group 1 element most likely to lose an electron
E   group 5 element most likely to grab and hold an extra electron

Date: _____   / 4

12

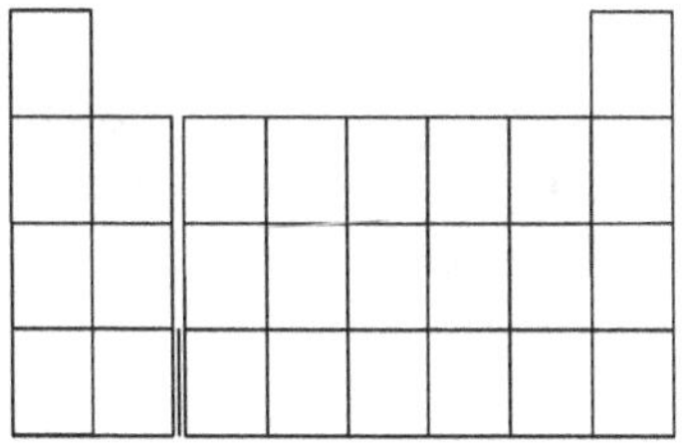

Mark boxes with appropriate letters to indicate:

A   a halogen smaller than chlorine
B   group 6 element bigger than sulfur
C   row 3 metal whose electrons are more loosely held than those of magnesium
D   group 1 element that is not a metal
E   row 3 element with a full valence shell

Date: _____   / 4

© Ross Lattner Publishing   www.rosslattner.ca

9 Academic Science
Student Exercises

Explaining Chemical Change

Lab 3.1: Reactions of Group 17, the Halogens

Do you Remember? Complete the Ross diagrams below.

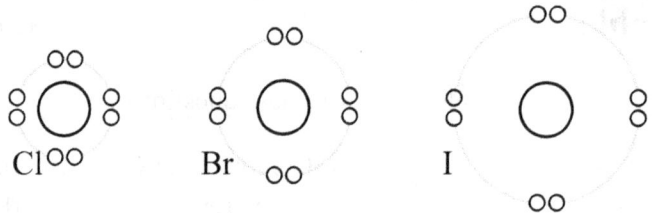

What's The Question? The Group 17 elements are fluorine, chlorine, bromine and iodine. They are also called the *halogens*. In this lab, you will examine some of their properties.

Which halogens are most and least reactive?

What Are We Doing?
1. Label three clean strips of paper I, Br and Cl. Your teacher will place one drop each of a solution of iodine (I_2), bromine (Br_2) and chlorine (Cl_2) on the paper. Gently waft some of the vapours toward your nose, and smell them. *Caution!* Describe the odour. Where else have you smelled things like that?
2. The teacher will prepare several substances to expose to the bleach.
3. **Predict** whether I_2 or Br_2 will be the strongest bleaching agent. **Explain** your prediction using core valence radius diagrams.
4. The teacher will expose the dyes to I_2, and then Br_2.
5. *Observe* how iodine and bromine react with the dyes. *Explain* your observations.

What Are We Thinking About?

Caution: Caustic Products. Cl_2, Br_2 and I_2 are dangerous to touch or inhale. *Never take your goggles off during this lab. Even if you finish, others may still be working.*

1. Would the force of attraction be stronger or weaker if the electron could get closer to the +7 core?

2. Which is the smallest halogen? Would its attraction for additional electrons be stronger or weaker than that of the other halogens?

Questions For Later...
1. Which halogen is most able to react with coloured dyes?

2. Which halogen has the strongest attraction for other electrons?

3. Explain why halogens are the most reactive non-metals.

Chemical Behaviour

Name:
Date:

Focus Question: Write the question that you are trying to answer.

1 *Predict:* Which halogen will be the most reactive bleaching agent... bromine, or iodine?

2 *Explain* your prediction, using both Ross diagrams, and sentences.

3 *Observe,* and record your observations here.

4 *Explain* your observations, using both Ross diagrams and sentences.

9 Academic Science
Student Exercises

Explaining Chemical Change

Lab 3.2: Properties of Group 1, the Alkali Metals
Do you Remember? Complete the Ross diagrams below.

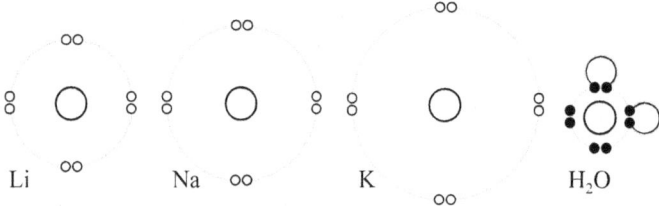

What's The Question? In the picture above, you can see which electrons are close to the +1 cores of the lithium, sodium and potassium atoms. You can also see the sizes and core charges of the hydrogen and oxygen atoms that make up water. Can water take an electron from an alkali metal? *From which alkali metal is the water most likely to take an electron?*

What Are We Doing?
Predict which metal is most reactive with water. *Explain* your prediction using diagrams like those above.

1. Put about 100 mL of fresh cold water in a 250 mL beaker. Dry the desk thoroughly.
2. Obtain a small (2 mm x 2 mm x 2 mm) piece of lithium in a petri dish. Cut it with a scoopula. Note its hardness, and the appearance of the fresh surface. Record.
3. Place the small piece of lithium in the water and immediately cover with the wire gauze. *Observe* from 1 m away. When the reaction ceases, record your observations.
4. Replace the water and repeat 2 – 3 with another metal.
5. Describe each element in the table below. Write words that are easy to compare.

Explain your observations, using core - valence - radius diagrams and complete sentences.

What Are We Thinking About?
1. **Caution: Danger of Fire, Caustic Products**. Li, Na and K are dangerous to touch, and react explosively with water. *Never take your goggles off during this lab: even if you finish, others may still be working.*

2. All of the atoms have the same charge on their atomic cores, and the same number of valence electrons. Why would one metal be more reactive than another?

3. What are the observable signs of a chemical change? Which visible signs are the most convincing to you?

Questions For Later...
1. In lab 3.1, the largest halogen atom was the least reactive. How do you explain the behaviour of the alkali metals?

2. The alkali metal below potassium in the Group 1 elements is cesium. Would cesium be more or less reactive with water than potassium?

Chemical Behaviour

Name:
Date:

Focus Question: Write the question that you are trying to answer.

1 *Predict:* Which alkali metal will be most reactive with water?

2 *Explain* your prediction, using Ross diagrams and sentences.

3 *Observe* the hardness and activity of the alkali metals. Record your observations here.

4 *Explain* your observations, using Ross diagrams and sentences.

9 Academic Science
Student Exercises

Explaining Chemical Change

Lab 3.3: Reactions of the Row Three Elements

Do you Remember? **Complete the Ross diagrams below.**

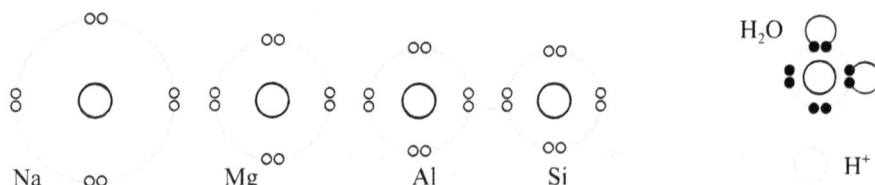

What's The Question? In the picture above, you can see how both the core charge and the atomic radius changes across a row of the periodic table. We will test the reactivity of each of these elements first with cold water, then with acid. (Acids contain the very small H⁺ particles shown above.)

How does the reactivity of the Row Three elements change left to right on the periodic table?

What Are We Doing?
Part A: Reactivity of elements with water.
1. Obtain small pieces of sodium, magnesium, aluminum and silicon.
2. Clean 4 small beakers, and put 20 mL of cold water into each one.
3. Put the sodium in the first beaker, magnesium in the second, aluminum in the third, and silicon in the fourth. Record your observations.
4. Pour out the water, but **keep the unreacted metals**.

Part B: Reactivity with Hydrochloric Acid
5. Put 10 mL of hydrochloric acid into each beaker that still contains unreacted elements.
6. *Don't* get more metal if it has already reacted with water in Part A.
7. Record your observations.
8. Clean up the beakers, and return any unused elements to the front bench.

What Are We Thinking About?
Caution: Danger of Fire, Caustic Products.
Sodium is dangerous to touch, and reacts explosively with water. Hydrochloric acid can injure eyes, skin, and damage clothing. *Never take your goggles off during this lab. Even if you finish, others may still be working.*

Among the elements Na, Mg, Al and Si above:
1. Which atom's electrons are easiest to remove?

2. Which atom's electrons are hardest to remove?

3. Which atom has the least ability to attract and hold other atoms' electrons?

Questions For Later...
1. Phosphorus is the element between silicon and sulfur. It is missing from this experiment. Look up phosphorus on the Internet or in an encyclopedia, and explain why we did not include it in this experiment.

2. Would you expect calcium to be more or less reactive than potassium? Explain.

© Ross Lattner Publishing   www.rosslattner.ca

Chemical Behaviour

Name:
Date:

Focus Question: Write the question that you are trying to answer.

1	*Predict* the order of activity with water, from most reactive to least reactive. Predict the order of activity with acid.	2	*Explain* your predictions, using the Ross diagrams and sentences.
3	*Observe* the reactivity of the row three elements with water and with acid, and record your observations here.	4	*Explain* your observations, using core - valence - radius diagrams, and sentences.

Questions For Later...(continued)

3. From which element(s) was water able to remove electrons? How do you know?

4. From which element(s) was hydrochloric acid able to remove electrons? How do you know?

© Ross Lattner Publishing   83   www.rosslattner.ca

9 Academic Science
Student Exercises

Explaining Chemical Change

Lab 3.4: Elements and Compounds

Do you Remember? List the 6 propositions of Dalton's Theory of Chemical Change

1. _____
2. _____
3. _____
4. _____
5. _____
6. _____

What's The Question?

Check out the radius, valence electrons, and core charge on oxygen and sulfur. Their small size and large +6 core charge suggests that they would attract and hold electrons from other atoms. Both have valence shells that have room for two more electrons. Do they form compounds?

What actually happens to the particles when two elements react to form a new compound?

What Are We Doing?

You will be given small samples of the elements copper, sulfur, magnesium, carbon, and oxygen in the air.

By gently heating pairs of elements together in a Bunsen burner flame, you will cause the elements to interact. **Do not overheat the elements. Heat them very briefly, just enough to start the reaction. Use as little heat as you possibly can.**

What Are We Thinking About?

1. What happens to the *speed* of particles when the temperature is raised?
2. Red hot is about 800 °C; white hot is about 2000 °C.
3. Air is about 18 % oxygen.
4. When two non-metals react, they both attract each others' electrons. This forms a *covalent compound*.
5. Non-metals take electrons away from metals to form an *ionic solid*.

Questions For Later...

1. Are *atoms* ever created or destroyed during a chemical change?

2. Are all of the *atoms* of the reactant still there after the product has formed?

3. List and describe the products of the metal + non-metal reactions.

4. List and describe the products of the non-metal + non-metal reactions.

Chemical Behaviour

Name:
Date:

After observing the reactions you must complete each diagram to include the following:
  a) Mark each of the three boxes either E *(Element)* or C *(Compound)*.
  b) Mark each box either A *(Atoms)* or M *(Molecules)*.
  c) Below each arrow, list three observable signs that chemical change has taken place.
  d) Draw particle diagrams to show the exact number of product molecules.

1. Place a tiny piece of sulfur on the insulated pad of the wire the gauze. Heat the gauze with the Bunsen burner for a few seconds, until the sulfur ignites *(Caution!! Do not overheat!)*. Remove the burner flame. Cautiously waft some of the gas produced toward your nose, and smell it.

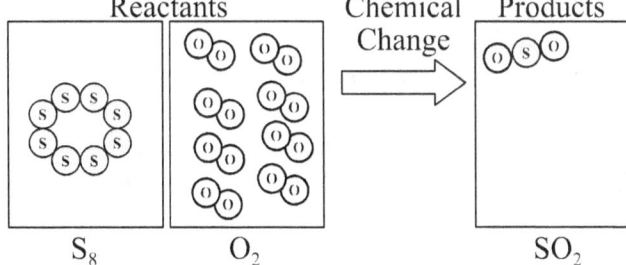

2. Place several small pieces of charcoal on the wire part of the gauze, and direct the burner flame onto the charcoal until it is red hot. Remove the flame, and observed the carbon carefully.

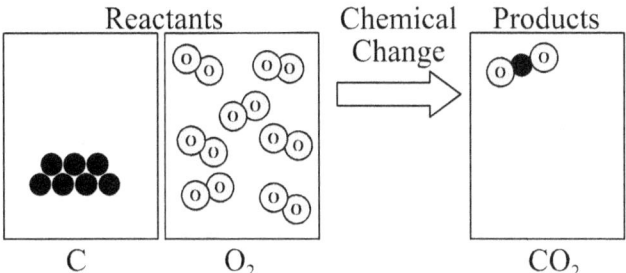

3. Hold a piece of magnesium ribbon in the tongs, and hold it in the burner flame until it ignites. Quickly move the burning magnesium over the steel sink.

   Do not look directly at the burning Magnesium. The intense heat can result in retinal damage.

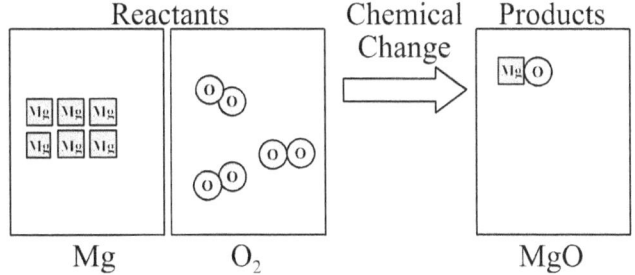

4. Place a small bit of sulfur on a clean penny. Using tongs, hold the penny in the flame for two seconds, or until the sulfur is just melted. Observe the formation of a new substance on the surface of the copper.

   Reheat for two seconds if necessary.

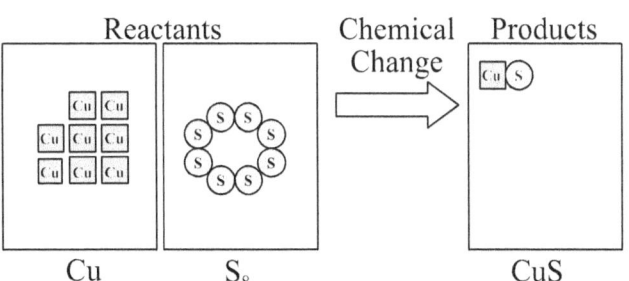

© Ross Lattner Publishing   85   www.rosslattner.ca

Explaining Chemical Change

Lab 3.5: Ionic Crystals, Covalent Molecules, and Networks

What's The Question? In this lab, you will test several solids for *odour*, *hardness*, and *melting point*. When you have a pattern of behaviour, you will decide whether two unknown compounds are ionic or covalent. *What are the differences between solids composed of ionic solids, and solids made of covalent molecules, and the covalent network solids?*

What Are We Doing? Clean the surface of the hot plate before you begin.

To test the melting point: Place a tiny crystal of each substance on a **cold** hot plate. You will have 6 different crystals on the hot plate at once. Plug the plate in, note the time at which each one melts. The longer the time, the higher the melting point.

To test hardness: crush one crystal between a scoopula and a glass plate. Judge hardness from the force with which the solid can be crushed.

To test odour: cautiously smell the mouth of the chemical jar by wafting a little air from the mouth of the jar toward your nose.

What Are We Thinking About?
1. *Ionic compounds* are formed from metals + non-metal. They form hard, gritty crystalline solids. (e.g. salt, chalk, baking soda, limestone, rust).
2. *Covalent compounds* are formed from non-metal + non-metal. These usually form small molecules. (e.g. water, sugar, oil, vinegar, wax).
3. *Covalent network compounds* are formed when non-metals atoms are linked into vast chains and networks. Usually really tough or hard. (e.g. diamonds, glass, quartz, mica, asbestos, nylon, hair, plastics).

Substance	Odour	Melting Point	Appearance	Hardness	Ionic or Covalent
Menthol					Covalent
Sodium Chloride					Ionic
Paraffin wax					Covalent
Magnesium Bromide					Ionic
Silicon					Network
Unknown A					
Unknown B					

Chemical Behaviour

Name:
Date:

Decide whether the following combination of elements will form ionic or covalent bonds. Then decide whether electrons will be transferred, or form shared pairs. Finally, draw the products as either an ionic crystal, a covalent molecule, or a covalent network solid.

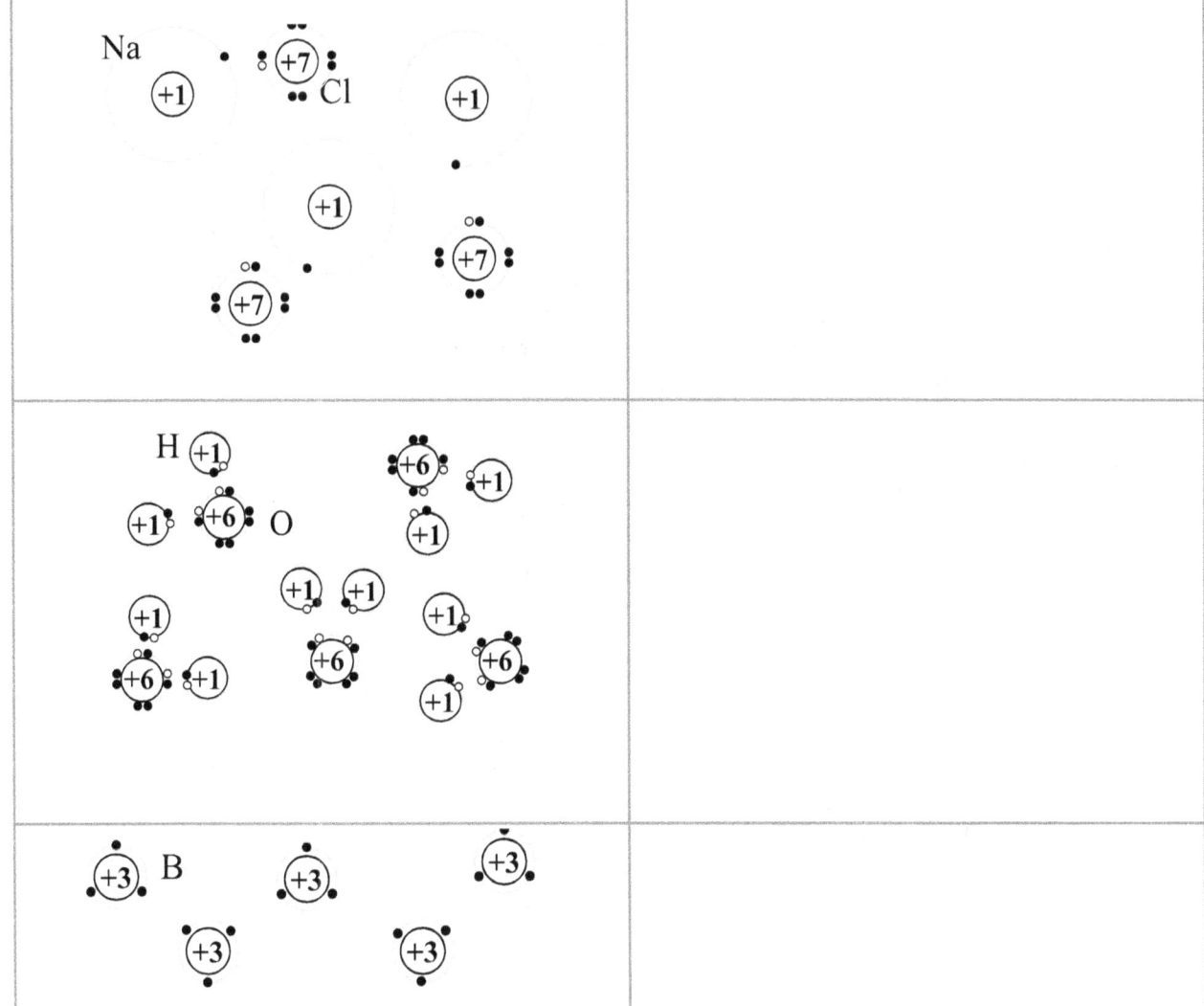

Questions For Later...

1. What generalization can you make about the melting point of ionic solids vs. covalent compounds?

2. What generalization can you make about the odour of ionic solids vs. covalent solids?

© Ross Lattner Publishing   87   www.rosslattner.ca

9 Academic Science Student Exercises

Explaining Chemical Change

Lab 3.6: Reaction of Acids with Metals

Do you Remember? List the 6 propositions of Dalton's Theory of Chemical Change.

1. _____
2. _____
3. _____
4. _____
5. _____
6. _____

What's The Question?
You often read about or hear about *acids*. They are all around you, even inside you. Acids can react with a large number of other substances, including metals. In this lab, you'll start a chemical reaction between hydrochloric acid and zinc metal, to produce a gas. You'll then use a chemical test to determine the identity of the gas. What is produced when an *acid* reacts with a *metal*?

What Are We Doing?
1. Clamp a test tube securely to the retort rod at a low angle. Add about 10 mL of acid.

2. Feel the temperature of the test tube of acid before the reaction begins, and after.

3. Gently slide one piece of zinc down the test tube into the acid. Place a one-holed stopper loosely into the mouth of the test tube.

4. Test the gas for both oxygen and hydrogen. Which gas is it?

5. Pour the liquid through a filter, and leave the solution in a Petrie dish overnight. Observe the residue in the next class.

What Are We Thinking About?
Caution: hydrochloric acid can injure eyes, skin, and damage clothing. *Never take your goggles off during this lab: even if you finish, others may still be working.*

2. The test for oxygen: Collect a test tube of the gas. Prepare a glowing splint. Thrust the glowing splint into the test tube of gas. If the ember bursts into flame, then the gas is rich in oxygen.

3. The test for hydrogen: Collect a test tube of the gas. Prepare a flaming splint Bring the flame to the mouth of the test tube. If the gas explodes, then the gas is likely hydrogen.

Complete the diagram with the correct number of each kind of particle. Label each box as containing either *Atoms (A)* or *Molecules (M)*. Then label each box as containing *Elements (E)* or *Compounds (C)*. Under the arrow list the observable signs of this chemical change.

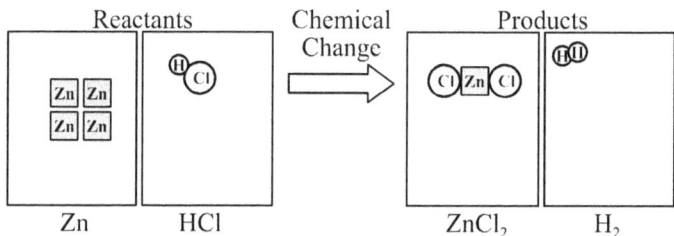

© Ross Lattner Publishing   www.rosslattner.ca

Chemical Behaviour

Name:
Date:

Focus Question: Write the question that you are trying to answer.

1 *Predict:* Is the gas produced in this reaction *hydrogen* or *oxygen*?

2 *Explain:* Tell why you believe your prediction.

3 *Observe,* and record your observations here

4 *Explain* your observations in a paragraph. Use particle diagrams and the Chemical Change Theory in your explanation.

Questions For Later...
1. What was the gas produced? Explain why you believe this.

2. From which reactant molecule did the gas molecules come? Explain.

3. During the reaction, the zinc metal disappeared. What happened to the zinc atoms?

9 Academic Science
Student Exercises

Explaining Chemical Change

Lab 3.7: Decomposition of a Covalent Molecule: Hydrogen Peroxide

Do you Remember? List the 6 propositions of the Chemical Change Theory

1. _____
2. _____
3. _____
4. _____
5. _____
6. _____

What's The Question?
"To decompose" means "to break down into simpler substances." Decompositions occur all around us, often in biological systems. One decomposition reaction you have already observed was the decomposition of water by electrolysis, Lab 1.5.

What are the products of the decomposition of hydrogen peroxide?

What Are We Doing?
1. Place a tiny (half a match head) quantity of manganese dioxide into the bottom of a dry test tube. Pour 10 mL of hydrogen peroxide into it. Place a one-holed rubber stopper LOOSELY into the mouth of the test tube.

2. Test the gas for oxygen and hydrogen, and record all of your observations.

3. After the reaction has stopped, examine the manganese dioxide. Carefully pour off any remaining liquid, and replace it with fresh hydrogen peroxide. Does the manganese dioxide still have all of its ability to decompose hydrogen peroxide?

What Are We Thinking About?
1. A catalyst is a substance which speeds up a reaction, but is not used up in the reaction.

2. What is hydrogen peroxide used for in your home?

Complete the diagram with the correct number of each kind of particle. Label each box as containing either *Atoms (A)* or *Molecules (M)*. Then label each box as containing *Elements (E)* or *Compounds (C)*. Under the arrow list the observable signs of this chemical change.

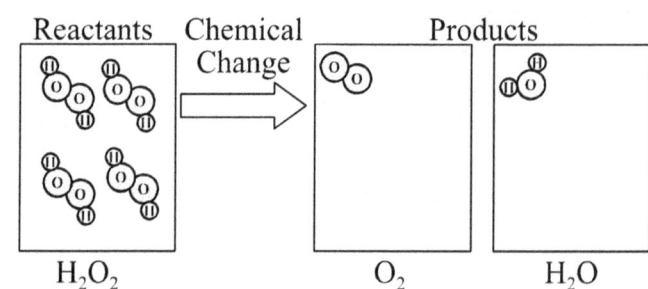

Chemical Behaviour

Name:
Date:

Focus Question: Write the question that you are trying to answer.

1 Predict: Is the product of the decomposition of hydrogen peroxide hydrogen, or oxygen?

2 Explain your prediction, using Dalton pictures and sentences.

3 Observe, and record your observations here.

4 Explain your observations, using Dalton's theory in diagrams and sentences.

Questions For Later...

1. After the reaction had stopped, and you replaced it with fresh hydrogen peroxide, did the manganese dioxide still have all of its power?

2. Was the manganese dioxide used up in the reaction?

3. What is the *catalyst* in this reaction?

4. Are the compounds in this lab *metals*, or *non-metals*? What kind of bonds hold them together?

Explaining Chemical Change

9 Academic Science
Student Exercises

Lab 3.8: Is All Carbon Dioxide the Same?

Do you Remember? List the 6 propositions of the Chemical Change Theory.

1. _____
2. _____
3. _____
4. _____
5. _____
6. _____

What's The Question?
Almost everyone has played with vinegar and baking soda at home. In this lab, we will be reacting baking soda (sodium bicarbonate) with vinegar (acetic acid) to make a gas. We will then compare the properties of that gas with gas from your breath, and gas from the combustion of charcoal. *What are the products when baking soda reacts with hydrochloric acid?*

What Are We Doing?

1. Place a heap of sodium bicarbonate about the size of your fingernail into a sandwich bag.
2. Put 5 mL of vinegar into a disposable plastic pipette. Without spilling a drop, put the pipette into the bag. Seal the bag.
3. Squeeze the pipette, allowing the acetic acid to mix with the sodium bicarbonate.
4. Test the gas.
5. Test your breath by bubbling your breath through some fresh limewater with a straw.
6. Test the gas produced when charcoal burns, to see if it contains the same gas.

What Are We Thinking About?

1. The test for carbon dioxide: Carefully and slowly, bubble some of the gas to be tested through a test tube of limewater.
2. Carbon dioxide is the only gas that makes limewater turn cloudy.
3. Carbon dioxide reacts with lime water to form calcium carbonate ($CaCO_3$), the same material in chalk, limestone, plaster, and marble.

$$CO_2 + Ca(OH)_2 \Rightarrow CaCO_3 + H_2O$$

Complete and label the diagram as before. Label each box either *Atoms (A)* or *Molecules (M)*, and as either *Elements (E)* or *Compounds (C)*. List the signs of this chemical change.

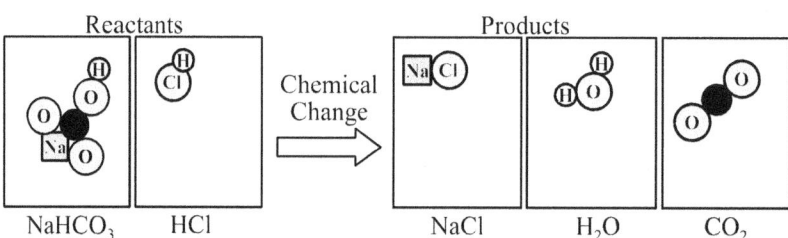

Chemical Behaviour

Name:
Date:

Focus Question: Write the question that you are trying to answer.

1 Predict: Will the carbon dioxide produced by your breath, combustion of charcoal, and reaction vinegar and baking soda all be the same?

2 Explain: Use the Dalton's theory of chemical change to give a reason for your prediction.

3 Observe, and record your observations here. Include both a description of the reaction itself, and the results of the test for carbon dioxide.

4 Explain your observations. Use the particle diagram on the previous page to help you think it through.

Questions For Later...

1. Is there carbon dioxide in your breath?

2. How does this experiment support or confound Dalton's Theory of Chemical Change?

3. If plants use carbon dioxide, does it matter whether it comes from people or cars?

9 Academic Science
Student Exercises

Explaining Chemical Change

Lab 3.9: Does Mass Change During a Chemical Reaction?

Do you Remember? List the 6 propositions of Dalton's Theory of Chemical Change.

1. _____
2. _____
3. _____
4. _____
5. _____
6. _____

What's The Question?
We have observed a number of chemical reactions, in which new substances were produced. Think back to some of the reactions. Did the products have more mass, less mass, or the same mass as the reactants?

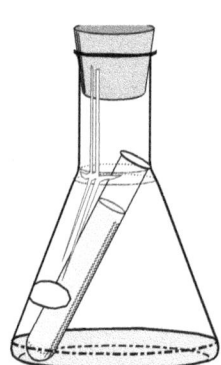

What Are We Doing?
1. Seal up a flask as shown, containing two different reactants.

2. Measure its mass as accurately as possible.

3. Mix the chemicals by tipping the flask.

4. Measure the mass again, to see if any new substances formed added or removed mass from the flask.

Questions For Later...
1. Did the mass change significantly during any of the reactions?

2. Would the mass have changed if the flask had not been sealed?

3. When you burn a tankful of gasoline in your car, does the mass of the gasoline disappear? Explain.

© Ross Lattner Publishing   www.rosslattner.ca

Chemical Behaviour

Name:
Date:

Focus Question: Write the question that you are trying to answer.

1 Predict: Will the mass of the flask increase, decrease, or remain the same during the chemical change?

2 Explain, using the Dalton's theory of chemical change in both pictures and a paragraph.

3 Observe, and record your observations here. Include both qualitative and quantitative observations.

4 Explain your observations, using both diagrams and Dalton's theory of chemical change.

© Ross Lattner Publishing   www.rosslattner.ca

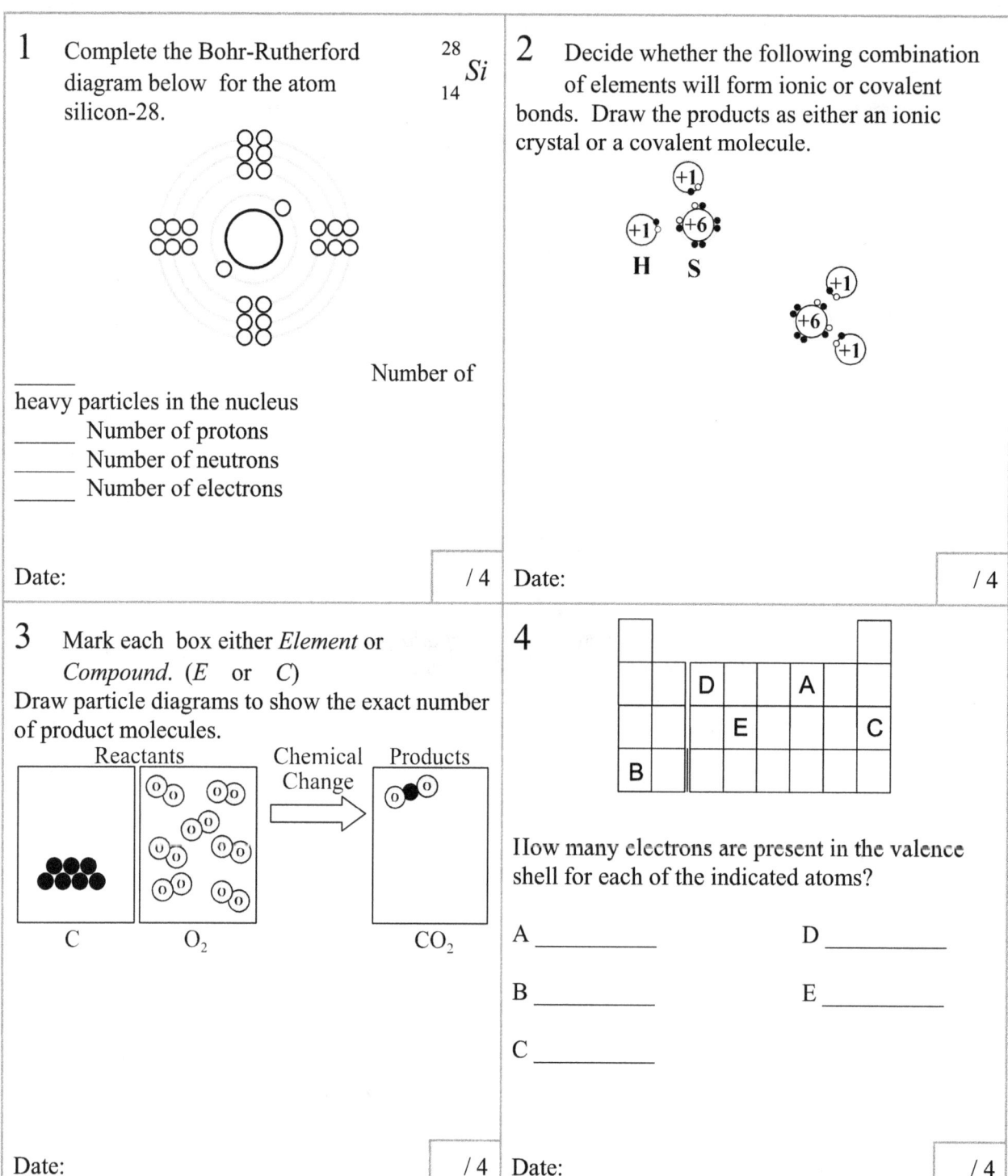

All the news that's fit to print... and then some
The Grade Nine Daily

Quiz 3.10:   Chemical Change   Name:

You may need to refer to a simple periodic table to answer these questions.

---

5   Complete the Bohr-Rutherford diagram below for the atom potassium-39. $^{39}_{19}K$

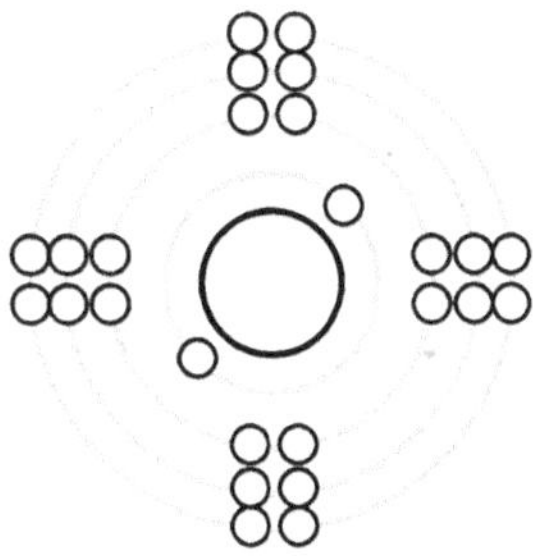

_____ Number of heavy particles in the nucleus
_____ Number of protons
_____ Number of neutrons
_____ Number of electrons

Date: / 4

---

6   Decide whether the following combination of elements will form ionic or covalent bonds. Draw the products as either an ionic crystal or a covalent molecule.

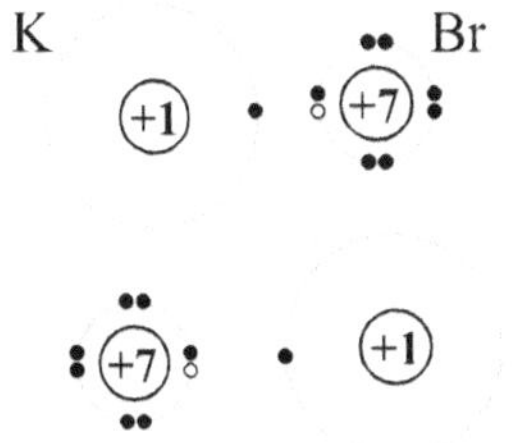

K   Br

Date: / 4

---

7   Mark each box either *Element* or *Compound*. (*E* or *C*)
Draw additional particles to make the mass of the reactants equal the mass of products.

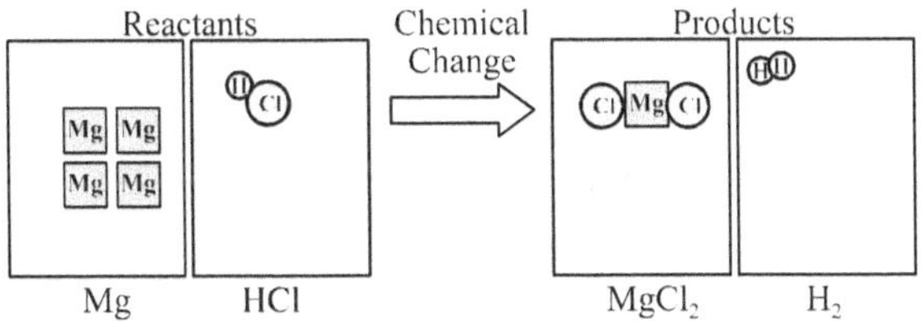

Reactants — Chemical Change — Products

Mg   HCl $MgCl_2$ H_2

Date: / 4

8

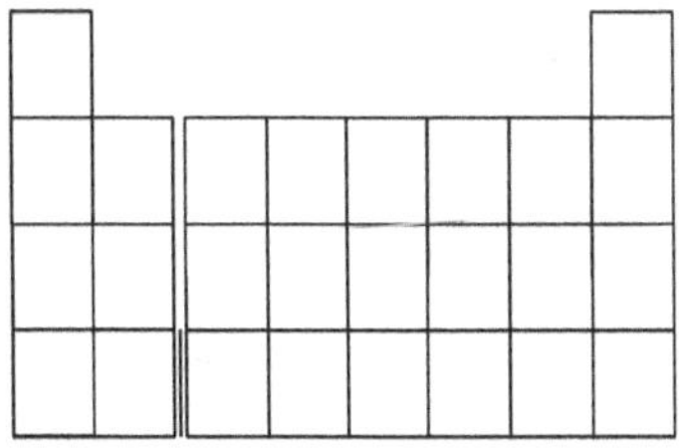

Mark all of the boxes with the appropriate letters to indicate:

A   all of the Noble Gases
B   all of the Alkali Metals
C   all of the Halogens
D   all of the Metalloids
E   all of the Alkali Earth metals

Date: / 4

© Ross Lattner Publishing   www.rosslattner.ca

The Grade Nine Daily

All the news that's fit to print... and then some

Quiz 3.10: Chemical Change   Name:

You may need to refer to a simple periodic table to answer these questions.

9

Mark boxes with appropriate letters to indicate:

A   a halogen larger than bromine
B   group 6 element smaller than sulfur
C   row 3 metal, most loosely held electrons
D   group 1 element that is *not* a metal
E   row 2 element with a full valence shell

Date: / 4

---

10 Decide whether the following combination of elements will form ionic or covalent bonds. Finally, draw the products as either an ionic crystal or a covalent molecule.

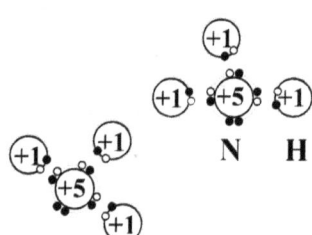

N   H

Date: / 4

11

Mark boxes with appropriate letters to indicate:

A   the element with only one electron
B   the smallest Noble Gas
C   most likely to grab and hold an extra electron
D   group 1 element most likely to lose an electron
E   group 5 element most likely to grab and hold an extra electron

Date: / 4

---

12 Mark each box either *Element* or *Compound*. (*E* or *C*)

Draw more particles to show the exact number of molecules in the balanced reaction.

Reactants   Chemical Change   Products

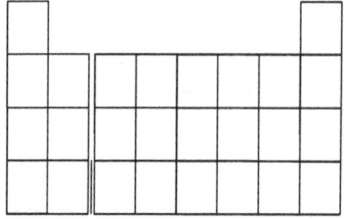

H_2 N_2 NH_3

Date: / 4

© Ross Lattner Publishing   www.rosslattner.ca

All the news that's fit to print... and then some

The Grade Nine Daily

Quiz 3.10: Chemical Change   Name:

You may need to refer to a simple periodic table to answer these questions.

---

13 Complete the Ross diagram below for the element sulfur. $^{32}_{16}S$

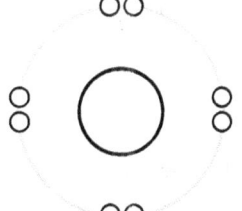

_____ positive charges in the atomic core
_____ number of valence electrons

Date: _____ / 4

---

14 Decide whether the following combination of elements will form ionic or covalent bonds. Finally, draw the products as either an ionic crystal or a covalent molecule.

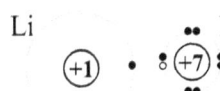

Date: _____ / 4

---

15 Mark each box either *Element* or *Compound*. (*E* or *C*)
Draw additional particles to make the number of atoms balance.

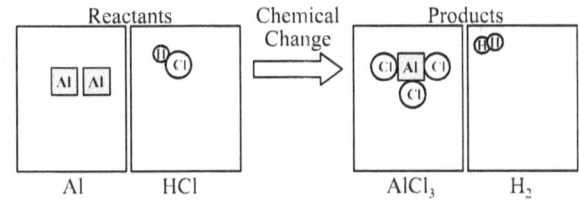

Date: _____ / 4

16

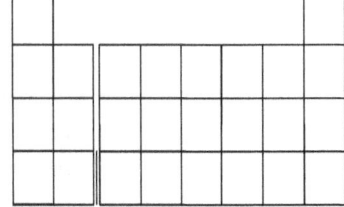

Mark boxes with appropriate letters to indicate:

A a halogen smaller than chlorine
B group 6 element bigger than sulfur
C row 3 metal whose electrons are more loosely held than those of magnesium
D group 1 element that is not a metal
E row 3 element with a full valence shell

Date: _____ / 4

Appendix: Laboratory Safety

9 Academic Science Student Exercises

The Hazards	*The Safe Way*
In this column is a list of lab safety issues that you will face in this course.	**Read this column to find out how to safely handle the laboratory problem.**
Eye Injury is possible from flying fragments of metal, glass or chemicals; from heat or flames; from caustic solutions such as acids or bases.	*Always wear safety glasses* in the laboratory. Never take your glasses off, even if you have finished your experiment. Other students may not have finished their lab work. The safety glass symbol indicates exercises in which safety glasses *must* be worn.
Crowding, Pushing and Horseplay increase the likelihood of a serious injury.	*Attend to your work.* Stay at the station you were assigned, so that there is room to work safely. If your teacher finds that your behaviour is a safety hazard, he or she may remove you from the lab. There is no place for behaviours which place others at risk of injury. Not at school, not at home and not at work.
Disorganized and Dirty Working Conditions are a hazard wherever they are found.	*Keep Lab Area Clean.* Clean and put away unused equipment. Tell your teacher about chipped, cracked, damaged or broken equipment. Do not leave anything on the floor, the desktop, the sink, or the cupboards that is not supposed to be there.
Broken Glass happens even to careful scientists.	*Do Not Touch* broken glass with your hands. Tell your teacher. When instructed to do so, use a broom to sweep the glass into a dustpan. Dispose of the broken glass in the special container provided. Do not leave it in the regular wastebasket: it could seriously injure a custodian.
Liquid Spills may consist of water, but they may also contain acids, bases, or toxic chemicals. You may not be able to tell the difference.	*Tell your teacher* about any spills immediately. Do not attempt to clean up without teacher instruction. Only if the teacher decides it's safe, use a cloth or paper towels to soak up excess liquid. Wipe the area clean with a damp cloth. Rinse the cloth frequently in fresh water. Wash your hands afterwards.
Solid Spills may consist of highly reactive chemicals. You may not know the specific hazards.	*Tell Your Teacher* about the spill, whether or not you caused it. Your teacher will instruct you on the safe way to handle the problem. In any case, the spill must be cleaned up promptly.

Appendix: Laboratory Safety

Name:
Date:

Open Flames are a frequent hazard. The Bunsen burner is the most likely safety hazard.	**Review Safe Handling of a Bunsen Burner** with your teacher. Be prepared to show how to light, operate and extinguish the burner at any time. Do not attempt to ignite pens, papers, rulers or other things. That kind of behaviour will certainly result in your being put out of the lab.
Fire. Any liquid solid or gaseous fuel burning where you do not want it to burn is a fire.	**Tell the teacher immediately!** Do not attempt to extinguish the fire with your hands, books, paper towels etc. Do not panic. Move away from the hazard. *Your teacher is the best judge of the appropriate course of action.*
Hot Metal or Glass cause more burns than any other hazard. There is usually no visible indication that they are hot. Glass in particular causes small, deep burns.	**Let Hot Objects Cool for 10 - 15 Minutes** before handling. Place all hot objects on a heat resistant pad. You and your partner will know where they are. Approach hot objects cautiously. Touch them at the coolest point first (the base of the retort rod, the bottom of the Bunsen burner or hot plate, the thumb screw of the iron ring). Use dry, not damp, paper towels to handle hot objects.
Hot Liquids such as boiling water or hot oil spread and splash rapidly. They also cling to skin and clothes.	**Let Hot Liquids Cool for 10 - 15 Minutes** before handling. Do not heat liquids in closed containers. Use hot plates rather than shaky retort rod assemblies. Do not heat more liquid than you need.
Obstructed Passageways prevent you from moving out the way of a spill or a fire.	**Stand at Your Lab Station.** Do not bring chairs or stools over to sit down. Your chair will prevent others from moving away from a spill or a fire.
Long Hair or Loose Clothing is more likely to become involved in your equipment. It can cause spills and breakage, or catch fire.	**Tie Back Long Hair; Secure Loose Clothing.** Outerwear in particular must be avoided in the lab situation. Jackets, sweat suits, hoods, etc are too large and awkward for the lab situation. They are also frequently made of materials that are flammable and can melt and stick to the skin in a fire. Avoid using laquer based hair sprays. A curly head of hair with hair spray can burn up completely in seconds.
Unauthorized Experiments can have unintended results.	**Stick to the plan.** Read instructions very carefully the night before the lab. Ask questions. Do not try experiments "just to see what happens." The dangers are too great.

www.ingramcontent.com/pod-product-compliance
Lightning Source LLC
Chambersburg PA
CBHW080446110426
42743CB00016B/3289